Risk and Exploration
EARTH, SEA AND THE STARS

Library of Congress Cataloging-in-Publication Data

Risk and exploration: Earth, sea and the stars, NASA administrator's symposium,
September 26-29, 2004, Naval Postgraduate School Monterey, California / Steven J. Dick
and Keith L. Cowing, editors.
    p. cm.

1. Scientific expeditions—Congresses. 2. Underwater exploration—Congresses. 3. Outer
space—Exploration—Congresses. 4. Technology—Risk assessment—Congresses. 5.
Science—Moral and ethical aspects—Congresses. I. Dick, Steven J. II. Cowing, Keith L.

Q115.R57 2005
910'.9--dc22   2005004470

Risk and Exploration
EARTH, SEA AND THE STARS

NASA Administrator's Symposium
September 26–29, 2004
Naval Postgraduate School
Monterey, California

STEVEN J. DICK AND
KEITH L. COWING, EDITORS

National Aeronautics and Space Administration

Office of External Relations
NASA History Division
Washington, DC

NASA SP-2005-4701

TWENTY YEARS FROM NOW YOU WILL BE MORE DISAPPOINTED

BY THE THINGS YOU DIDN'T DO THAN BY THE ONES THAT YOU DID DO.

SO THROW OFF THE BOWLINES. SAIL AWAY FROM THE SAFE HARBOR.

CATCH THE TRADE WINDS IN YOUR SAILS. EXPLORE. DREAM. DISCOVER.

Attributed to Mark Twain

Contents

SESSION ONE—EARTH

SESSION TWO—SEA

SESSION THREE — THE STARS

SESSION FOUR — WHY WE EXPLORE

APPENDIX — IN CLOSING

Acknowledgments

The NASA History Division is pleased to present the record of a unique meeting on risk and exploration held under the auspices of the NASA Administrator, Sean O'Keefe, at the Naval Postgraduate School in Monterey, California, from September 26–29, 2004. The meeting was the brainchild of Keith Cowing and astronaut John Grunsfeld, NASA's chief scientist at the time. Its goals, stated in the letter of invitation published herein, were precipitated by the ongoing dialogue on risk and exploration in the wake of the *Columbia* Shuttle accident, the Hubble Space Telescope servicing question, and, in a broader sense, by the many NASA programs that inevitably involve a balance between risk and forward-looking exploration. The meeting, extraordinarily broad in scope and participant experience, offers insights on why we explore, how to balance risk and exploration, how different groups define and perceive risk differently, and the importance of exploration to a creative society.

At NASA Headquarters, Bob Jacobs, Trish Pengra, and Joanna Adamus of NASA Public Affairs led the meeting's implementation. At NASA's Ames Research Center, Director Scott Hubbard coordinated a group including Rho Christensen, Danny Thompson, Shirley Berthold, Victoria Steiner, Ed Schilling, Mike Mewhinney, Kathleen Burton, and the Ames Video Team. Mel Averner also contributed significantly to the concept and content. The Naval Postgraduate School, commanded by Rear Admiral Patrick W. Dunne, provided a congenial venue.

The meeting was broadcast on NASA TV, and thanks are due in this regard to Al Feinberg, Tony Stewart, Jim Taylor, and the planners' collaborative: Mark Shaddock and Spotlight Productions, Donovan Gates of Donovan Gates Production, and Michael Ditertay and his staff on this 30-person television crew. Thanks to their efforts, a DVD record of the meeting has also been produced.

Thanks are also due to the moderators: Miles O'Brien of CNN, Chris McKay of NASA Ames, David Halpern of the White House Office of Science and Technology Policy, and John Grunsfeld, NASA Headquarters.

Finally, thanks are due in the NASA History Division to Liz Suckow and Steve Garber, and to the staff in the Printing and Design Office at NASA Headquarters: Shelley Kilmer, Cathy Wilson, Paula Kephart, Henry Spencer, Jeffrey McLean, Greg Treese, and Steve Johnson.

In order to maintain the informal flavor of the meetings, these proceedings are based on transcripts that have been lightly edited for grammar and punctuation. Most references to slides shown during the presentations have been deleted.

Steven J. Dick, NASA Chief Historian
Keith L. Cowing, Editor, NASAWatch.com

Invitation Letter

The goal of the Risk and Exploration Symposium is to engage in an open discussion about the issue of risk—identifying it, mitigating it, accepting it—all in the course of exploration. Yes, risk taking is inherently failure-prone. Otherwise, it would be called "sure-thing-taking."

Challenge fosters excellence, often drawing on previously untapped skills and abilities. Each of us takes and accepts risk as a part of our daily existence. We often go out of our way to seek challenge. However, seeking challenge often means accepting a high level of risk. The dictionary defines risk as being exposed to hazard or danger. To accept risk is to accept possible loss or injury, even death.

One of the key issues that continues to be debated in the tragedy of the Space Shuttle *Columbia* is the level of risk NASA accepted. And, ultimately, the entire nation is now engaged in a broader debate over whether or not the exploration of space is worth the risk of human life.

While risk can often be reduced or controlled, there comes a point when the removal of all risk is either impossible or so impractical that it completely undermines the very nature of what NASA was created to do—to pioneer the future.

Everyone today understands that human space exploration is a risky endeavor. However, the quest for discovery and knowledge, and the risks involved in overcoming seemingly insurmountable obstacles is not unique to NASA. Whether the challenge is exploring the depths of our oceans or reaching the top of our highest mountains, great feats usually involve great risk.

During this symposium, we want to examine the similarities between space exploration and other terrestrial expeditions, and examine how society accepts risk. For example, more than 40,000 Americans die each year in automobile accidents. A recent study of 22,000 fatal accidents showed that nearly two-thirds of the victims were not wearing seatbelts—a clear indication that too many of us fail to understand the risks when we get behind the wheel of a car and fail to buckle up.

Why are sacrifices made in the name of exploration more notable than the losses suffered in the course of everyday life? What lessons can be learned by studying the history of exploration and risk? And why are so many people willing to risk their lives to advance adventure, discovery, and science when, often, the benefits are unknown and indefinable?

We have assembled an invitation-only audience of participants for this important event comprised of NASA astronauts and leaders, as well as world renowned mountain climbers, deep sea explorers, cave explorers, Arctic and Antarctic researchers, scientists, communication experts, and others. These participants are involved, in a personal way, with risky endeavors, which serve to expand the frontiers of human knowledge beneath the sea, on the surface of Earth, and in outer space.

We look forward to your participation in what we believe will be a spirited and highly beneficial public discussion of risk and exploration.

Sean O'Keefe, NASA Administrator

Introduction

On behalf of Ames Research Center, one of your local hosts, I bid you greetings: to the Administrator, to my fellow Center Directors, distinguished guests from Headquarters, members of the NASA family. Our thoughts go out today to members of the Kennedy Space Center who cannot be here with us. They are attending to the damage from the hurricanes that have hit them recently. Greetings also to the rest of the distinguished guests here, and, of course, to our gracious hosts at the Naval Postgraduate School.

We are here today to look at risk. How do we perceive risk, real and otherwise? How do we identify risk? How do we mitigate risk? And finally, when and how should we tolerate and accept risk? And all of this discussion, of course, is in the context of exploration, the essence of what we do.

What is risk? In our daily lives, we all assume multiple levels of real risk. We get into cars. We get into airplanes, some of us far too often. We cross roads. We use cell phones while we're driving. We eat bacon and barbecued chicken that contain potentially carcinogenic nitrites. One way or another, all these things represent real risks that we tolerate and accept. Most of us mitigate these risks in some measure by doing things like wearing seatbelts, using the oversight of agencies like the FAA, remembering to look both ways, and getting our annual physical checkups. We are accustomed to living with risk, whether we realize it or not.

Today, tomorrow, and Wednesday morning, we are going to explore risk in the field of exploration. There is monetary risk, programmatic risk, and, of course, the far more compelling issue of risk to life and limb of the explorers. We look at how we decide what risks are acceptable, both for the individual and the institution. I think this is a key element here. How do people perceive risk individually? How do we perceive risk as institutions? How do we collectively decide what we will accept?

Let me give you a little bit of insight from my own experiences. I will start with a program from seven or eight years ago where I was fortunate to be the manager of the Lunar Prospector mission. In December 1997, just days before the launch, we had to make some very tough assessments of the risk. There was a single-string spacecraft, first use of a new launchpad, the maiden flight of a new launch vehicle, although it was a design based on some proven fleet ballistic missile motors. There was a very tight budget schedule, and a relatively young team, although with some key experienced people at the top. So, why did we go for launch? Why did we accept that risk?

Well, we had a strong test program, solid teaming with effective communication and some very experienced key managers, open channels during the design and development process, and lots of insight from a "graybeard" committee that provided us with truly effective comments. In the end, the deciding factors amounted to a considerable degree of trust and the determination that we had done everything possible to ensure mission success. So, we launched. That mission was successful, and that data is now being used as we plan a return to the Moon in the not-too-distant future.

Another example is from the restructuring of the Mars program. In April and June 2000, we had to decide whether to take the risk of going back to Mars for an opportunity that was just 38 months away. Would we not only send an orbiter, but would we send a lander? We had three opportunities there: do nothing, send an orbiter, and send a lander. Why did we take the risk of doing the most ambitious one?

Well, it was an outstanding launch opportunity. We had solid heritage from the Pathfinder Mission. We had a fully developed payload, and we had a robust entry, descent and landing scheme that was well understood. We put two rovers into the mission to balance the risk against random failures, and I think, as you will hear later on in this meeting from Steve Squyres, having 2 rovers had an unanticipated benefit of providing a rich panoply of hardware from which to reduce other types of risks. But, in the end, the reason we took that risk was because it was the right thing to do. A strategic plan brought us back to the surface of Mars and gained the kind of information that is now changing the textbooks, changing the way that we look at the Red Planet.

Finally, let's look at the issue of human spaceflight risk. I was honored to be the sole NASA representative on the Columbia Accident Investigation Board. It was one of the most difficult assignments that I have ever had in my career. In that entire seven months, we never lost sight once of the nine lives that were lost, the seven astronauts and the two searchers, the two helicopter pilots. Clearly, we take substantial risk when we put humans on the top of a rocket and leave Earth's gravity well. How do we adequately mitigate that risk? From *Columbia*, we learned some tough lessons. We learned we need to develop a culture of safety for the long haul. We need clear communication, clear organization, adequate resources, rigorous and sound engineering principles, and a program systems engineering approach that addresses the entire effort, that addresses all the analysis that we need.

We also need to effectively learn to encourage alternative points of view. If we do all of these things, can we eliminate risk, especially for human spaceflight? I think not. Will we find a balance of well-mitigated risk and a powerful level of acceptable risk? I, for one, think so. What will that balance be? That's what we have come together to discuss today.

Scott Hubbard, Director, NASA Ames Research Center

The Vision for Space Exploration

There is no historic antecdnt I think, to a gathering like this. Certainly, no records exist of people living in Lisbon 500 years ago attending a candlelight symposium featuring Amerigo Vespucci or Vasco da Gama or Ferdinand Magellan. So, this is an opportunity given by modern technology and the ease of transportation to pull together this really extraordinary group of folks who've experienced the full extent and breadth of exploration and the risks attendant thereto. Such a gathering was important for the purpose of parsing this larger question of risk and return on the exploration ventures we are about. I am particularly grateful to John Grunsfeld, who has really provided the intellectual horsepower behind this kind of effort to think about these questions in a structured way, and to Keith Cowing, two very disparate kinds of folks, but folks who share the passion and desire for exploration and an understanding of the attendant risk to it. So, to Keith and to John, I am most grateful for that extraordinary nudge that you all provided in pulling this together and providing the structure of the meeting.

We are gathered here, appropriately, in a place like Monterey, at the edge of a great ocean, to discuss exploration in all of it facets of extreme environments here on Earth and in space. Indeed, this historic location is steeped in a history of exploration. The ventures of

Sean O'Keefe
NASA Administrator, 2001–05

Sean O'Keefe was appointed by President George W. Bush to serve as NASA's 10th administrator. He served from 21 December 2001 to 19 February 2005. During his term, O'Keefe actively supported NASA's many missions to advance exploration and discovery in aeronautics and space technologies.

O'Keefe joined the Bush Administration on inauguration day and first served as the Deputy Director of the Office of Management and Budget until December 2001. Prior to joining the Bush Administration, O'Keefe was the Louis A. Bantle Professor of Business and Government Policy, an endowed chair at the Syracuse University Maxwell School of Citizenship and Public Affairs. Appointed as the Secretary of the Navy in July 1992 by President George Bush, O'Keefe previously served as Comptroller and Chief Financial Officer of the Department of Defense from 1989. Before joining then Defense Secretary Dick Cheney's Pentagon management team in these capacities, he served on the United States Senate Committee on Appropriations staff for eight years, and was Staff Director of the Defense Appropriations Subcommittee. His public service began in 1978 upon selection as a Presidential Management Intern.

so many people to explore and to establish the site of civilization that we see in this marvelous area here around Monterey is testimonial to that. What we enjoy each day in this community, and understand about exploration and its benefits, are here and evident each day. Certainly this evening we will have an opportunity to see that more specifically at the aquarium.

I want to provoke some thought and reflection about a central question which we're discussing here in these two and a half days. Why do we take such risks to explore? As humans, what is it about us that really wants to understand that which is on the other side of the horizon, that which is on the other side of the ridge? In doing so, there are periods of our human history in which the acceptance of those risks have resulted in great gains and, in other cases a mere footnote, because it ended in a way that was less than fulfilling. In each case, there was always a contribution to that human desire to want to know and understand. How we assess those risks and deal with the challenges of exploration is the central question we are about in this two and a half days, and I am most grateful to all of you for accepting the invitation to participate in this kind of debate and discussion of how we may structure this question, not only in a public but also in a specific way. I am certain we will have a lively discussion of where you draw the line between the benefits of exploration and the inherent risks, especially as technology changes, and as we learn more about the environments in which we explore.

Now, this is in part about NASA participation, to be sure, but it is mostly about those of us from NASA having the opportunity to learn from so many others who are engaged in the broader exploration agenda of the central questions we pose.

I am most grateful to see the Apollo, Shuttle, and Space Station veterans who have gathered here with us to share their thoughts. Indeed, I think it will be historic in and of itself to learn so much from them. All of them have dared to sit in a spaceship at one point (and in several cases, like Jerry Ross, seven different times), to sit on the top of the spaceship with millions of pounds of explosive fuel, prepared to put their lives on the line in order to advance that cause of exploration and discovery. Now I asked Jerry, why you do this, and he said, "Well, because it's an opportunity to do so," and he would easily sign up for an eighth flight this afternoon, I'm sure. As a matter of fact, I don't think he would wait until noon to sign up for an opportunity.

To some, it may seem that NASA has made space travel routine. Let there be no mistake: I think we all fully appreciate and understand that space flight and exploration is still a very risky proposition. Despite our efforts to eliminate that risk, there will always be an attendant risk to such a venture. And, as a result, here in attendance are NASA scientists, engineers, and managers whose job it is to have constant vigilance about that risk. And in that regard, I view myself as included in that requirement for constant diligence to assure that risk is mitigated as much as we can.

From the discussions that will take place here, I hope we will gain a greater appreciation of our responsibility as a public organization to take on bold and risky ventures, and to learn from those who have accepted private ventures and other approaches to how we explore risk. We want to know how to frame that discussion and debate, and evaluate that risk in a different way.

But, again, it is also a requirement that we do that in a diligent manner that minimizes and mitigates, to the maximum extent we can, what that risk may be, that we understand what risk is as much as possible and, in some cases, accept it relative to the returns we think are feasible. That's the price of admission of what we do each and every time we're engaged in any exploration venture, be it of human spaceflight or robotic probes. It is always measured in the public domain

WHY DO WE TAKE SUCH RISKS TO EXPLORE? AS HUMANS, WHAT IS IT ABOUT US

THAT REALLY WANTS TO UNDERSTAND THAT WHICH IS ON THE OTHER SIDE OF

THE HORIZON, THAT WHICH IS ON THE OTHER SIDE OF THE RIDGE?

and in the public eye relative to what our expectations are to that return. Indeed, NASA is an agency that has been defined over the course of its 46 years by great, great triumph and unbelievably deep tragedy, and we've learned from both ends of that spectrum. It's a consistent set of themes. It is, indeed, the singular aspect of what has described this agency throughout the course of its four decades.

We have purposely expanded the list of invitees, and we are very grateful to the folks who have accepted to be a part of this. We want to gain an added perspective from the people engaged in exploration of the Earth's most extreme environments, and learn what they can bring to the question of why we explore in the face of danger. What is it about that act of exploration that makes it so appealing? And so important? And so much of an acceptance of human desire to want to understand and know that which we don't?

Within the NASA family, we have great respect for all who put their lives on the line: Not just to seek thrills, but, rather, to gain knowledge, wisdom, and experience that will benefit all humanity. All of those assembled here have a unique and exciting story to tell about what drives us to explore, whether engaged in it directly or specifically involved in supporting its effort, all with the same objective.

All of those stories, I have no doubt, we'll hear at this meeting. Also, we'll learn from the experiences of how folks work to minimize and mitigate the risk, and learn where the fine line is between responsible and imprudent risk. Where is the differentiating line that marks that? Even when we've applied a careful calculus to these kinds of circumstances, in many cases and in many circumstances, the events of nature will provide a set of risks that must be responded to, and challenges independent of whatever control we might have over it.

Our colleagues at the Kennedy Space Center, after their second hurricane in the span of a few weeks, are dealing with just that set of challenges, of risks that they are working through. And because of their extraordinary diligence, having survived two unbelievable events of what are natural disasters in their own right, nonetheless, have survived those experiences with all the Shuttle orbiters intact, all the space station hardware in great condition, and no loss of life, no injuries. It's an extraordinary testimonial to the amazing diligence of Jim Kennedy, the director of the Kennedy Space Center, and what the Kennedy team has done to ride out this set of natural disasters.

I was down at the Kennedy Space Center with Bill Readdy a week ago, and the poetic kind of discrimination with which nature provides us a set of challenges on risk were evident to us. I got an opportunity to see the Vertical Assembly Building, which is the dominant structure on the skyline of the Kennedy Space Center that all recognize, and you could literally tell which way the wind was blowing when Hurricane Frances blew in. Three of the four sides of the Vehicle Assembly Building were in relatively good shape. On one side of it, though, better than a thousand panels were blown off. Several of those panels have also departed as a result of the latest hurricane that just came through. As a consequence of striking some of the buildings in the area, ripping off big chunks of roof, all manner of consequence and destruction that occurred as a result of that, all of which was mitigated in some way, shape, or form. And yet, the irony is that right next to the Thermal Protection Building, where a portion of the roof blew off next to the Vehicle Assembly Building, there was a pressure-treated lumber gazebo without a scratch.

Nature discriminates very profoundly, and why it does, we don't understand. It certainly is a case that reminds all of us, even in such a simple example as that one, that despite our best efforts, there are unknowns that will always rise up in any of these circumstances, in any case of exploration, for which the only defense we have is diligence and the hope that we have mitigated against it as well as we can.

We're living in an era of great potential, one in which the exploration of the solar system and of the Earth's most extreme environments will boost the opportunities we have to become a smarter, safer, healthier, and more intelligent world. Certainly, we're more informed about the neighborhood we live in, a neighborhood defined as this little, bitty solar system around this little, puny star in a gigantic galaxy that is part of a massive universe. We are just on the cusp of understanding what our role is in that broader case, and it's only been in the last 40 years that we have come to understand it in ways that are really quite profound. I'm confident that if we do this right, we'll be amazed by the rapid pace of progress our future exploration activities will bring about.

But we also know from history about the consequences of forsaking exploration. When we evaluate and determine as individuals, or collections of people and nations, to forsake those exploration opportunities, it has consequences. In the 15th century, China had the opportunity to be the world's foremost maritime power and, indeed, possessed that capability. The Chinese ruling class, nonetheless, decided that the sponsorship of the fleet was an indulgence. History, in the course of the several centuries thereafter for that culture, is certainly a function of those choices that were made.

Certainly, we have the same opportunity in this country to make similar kinds of choices. In the 1875 time frame, the director of the Patent Office advised the President of the United States that it was a good time to close down the Patent Office, because everything that needed to be invented had been. Had the President of the United States accepted that wizened sage's advice at the time, imagine where we'd be! Yet, that was based on a calculated understanding of what folks thought was the potential of new inventions. It wasn't reached whimsically, it was reached by those who really believed that we had already incurred an enormous evolution of change of technology, revolution in industrial affairs, and, as a consequence, we were on a roll, and anything beyond that was going to be simply derivatives of the same.

In the last century, we've seen an explosion of growth in the exploration of seas, remote regions of the Earth, and, indeed, space. All of which, arguably, might not have happened had that original set of recommendations been followed.

It is no accident that NASA's founding occurred some 46 years ago this very week, in the same decade that Edmund Hillary and Tenzing Norgay first stood on our planet's highest peak, and that Jacques Cousteau used the good ship *Calypso* to conduct his epic voyages of undersea exploration. As explorers, we all share that common bond.

We dare to dream grand dreams, and, in the process of doing so, assume tremendous risk, some of it beyond the scope of our knowledge of the time in which those dreams are assumed and accepted. We do so for what we know to be great purposes. We also, in the depths of those tragedies that occur, grieve when our brethren are lost in the cause of exploration. Indeed, part of the impetus for this symposium was brought about in debates that occurred in the aftermath of the *Columbia* tragedy. It was a tough report that the Columbia Accident Investigation Board released. It told us an awful lot about the technical problems that led to it, the engineering challenges that we did not understand, and, as a result, paid an ultimate price with nine people—the seven members of the crew as well as two engaged in the recovery of *Columbia* after its destruction. We learned that that is a horrendous price—again.

But it also brought about, as a consequence of debate, a discussion about how we contributed to that tragedy, and a broader public debate about a renewal of the purpose of why we explore. And that debate has gone on in a broader public policy sense. The year after that horrific tragedy, it nonetheless was an impetus for motivating a debate by answering the fundamental question of why we explore and what the strategy and path ahead should be in pursuit of that human desire to understand.

In the process, it also raised a series of questions that we have the opportunity here, over this couple of days, to at least debate how they should be framed. We have, I think, as a consequence of the strategies the President has levied, and the direction that he has provided to us at NASA for exploration, a better understanding now of exactly how to pursue those exploration goals. And it's laid out in a series of objectives and programs to achieve it, and a stepping-stone approach, and a whole range of different ways in which we're going to achieve that task.

But communicating the *why* of this venture has just begun as a public debate in the last few months. Again, this is an extraordinary moment in time, in which there has been a renewal of that spirit of discovery and exploration. In part, it must then engage in this broader public dialogue, because we are, after all, a public organization for which there is trust that is rendered to us by the public for our acceptance of these kinds of challenges. And that trust is fragile, and at each of the intervals in which we have seen either those great triumphs or great tragedies, it has been tested.

So understanding the why, and being able to communicate that in a way that's effective, is part of what this discussion is all about. And while participating in the panel discussions, I would ask that each of us pose the following kinds of questions: How do we integrate the risk calculation with the benefits to be derived? What's the return? How do we communicate that as well?

Because it's apparent, when tragedies occur, what the depth of the risk was that was accepted, and then, therefore, not responded to effectively. But understanding what the benefits were to be derived sometimes gets lost in the translation, so how do we integrate that better? And that's on a personal as well as a societal level. There are any number of colleagues here, and those who've elected and chosen to participate in this venture, who can articulate this on a personal level. But, also, how we translate that in a broader societal context, I think, is very important, why we've accepted those risks, for what potential gain.

Also, ask the question: How do we regularly remind ourselves of the risk, and is that really important? Is it something we really need to focus on, and to what level of depth and degree? Certainly, being accepting of it or dismissive of it is not one of the options, but what is the appropriate balance? Also, pose the question: How do we avoid complacency? It is human nature, it is part of our human makeup, that what we see repetitively we begin to accept as normal? If you've never seen it before, it suddenly becomes a remarkable circumstance, something which you respond to because you've never seen it before. And, yet, it may be far less significant as risk than what you see every single day, yet, because we see it so regularly, we accept it.

What is it about our view as a culture, as a society, of why it is understood that there is a risk attendant to driving an automobile, flying in a commercial airplane? These are things we understand as being part of that, either intuitively or intellectually, and have recognized that despite the fact that lots and lots of folks every single year die in horrific automobile accidents, we accept that, as humans, because of the transportation and opportunities it provides, the facilitation of discourse and communication between and among each other, and the means to get from here to there. What is it about it that makes that an accepted level of risk?

And, yet, in the act of exploration, when the tragedies occur, what is it that makes that either intolerable or why we question it? And again, the root of this may yet well be grounded in how well we understand the benefit that we think we gained as a consequence of the activity and the effect of accepting that risk.

Also, for those who are involved in wider-ranging sets of exploration opportunities, what is it about the risk that you accept that's different than that which NASA accepts in what we do, and what is similar? How do you parse

. . . WE HAVE AN OPPORTUNITY TO LEARN FROM EACH OTHER'S EXPERIENCE SO THAT WE CAN GO FORWARD BOLDLY INTO THE UNKNOWN, INFORMED BY A RESPONSIBLE SENSE OF HOW WE COMMUNICATE IN A WAY THAT CONVEYS THE REASONS WHY IT IS OR IS NOT ACCEPTED AS AN APPROPRIATE LEVEL OF RISK.

between both and determine what we can learn from this about that? And, I guess, the ultimate question: What can we learn from each other by how to frame this question differently and, indeed, communicate it more effectively as an opportunity for great gain?

Over the course of human history every major advance has occurred because of the temerity on the part of human beings to want to understand and to explore and to do something that has not been tried or has been tried so irregularly as to have no pattern to it. If you think of every major advance in the course of our existence it has been attributed to that characteristic of us as human beings.

This week we have an opportunity to learn from each other's experience so that we can go forward boldly into the unknown, informed by a responsible sense of how we communicate in a way that conveys the reasons why it is or is not accepted as an appropriate level of risk. We are resolved at NASA to better communicate with the public about why it's necessary to take those risks, and why it is inherent in the way we, as human beings, conduct our lives in a way that would give meaning and purpose to this larger exploration agenda. At its core the answer is best summarized by a comment President Bush made in Houston just days after the *Columbia* tragedy, that this cause of exploration is not an option we choose, it is a desire written in the human heart. And when we can confront that, even on both ends of the equation—in its great triumph as well as in its depths of tragedy—and we're reminded why we're driven to this, what is it we can do responsibly as public servants, for those of us at NASA and in the broader community of explorers represented here, to communicate that more effectively?

I thank you all for your participation, and I look forward to sharing with all of you the spirit of exploration and discovery that I think is certainly evident in this group by so many people who have elected to spend their time to engage in these important questions. The manner in which we have framed this over these couple of days will bring those kinds of questions to bear in ways that, as we move forward in this next step of exploration, to return to flight, to complete the International Space Station, to develop through Project Constellation an opportunity to explore beyond Earth's orbit, all of this may be the beginnings again of an opportunity to frame that discussion and debate, not only among ourselves, but in the broader public, in ways that highlight those purposes of exploration, and why we engage in the risks and accept them, knowingly, for the purposes for which NASA began.

NASA Image/Renee Bouchard

Race to the Moon

a lot more modern astronauts who have done a lot more than I have ever done. But I think the previous speakers have really set the scene for this discussion of how we perceive risk. Now, I would like to expand this concept of risk as it pertains to spaceflight, and, of course, Apollo 13. But before I do, let me digress and tell several personal stories familiar to me of how I think risk is perceived.

The first story takes place long before we had NASA astronauts. After World War II, Wernher von Braun came over from Germany, and he and his team went out to White Sands, New Mexico with a bunch of dilapidated V-2 rockets. Their job was to fire those rockets up into the upper atmosphere and, with the proper sensors, determine maybe what the stratosphere was like—the flow, the elements, and things like that.

But von Braun was a very farseeing individual. He knew that someday man would go into space, and he would piggyback on these rockets some experiments that would determine, or help to determine, if man could survive in the environment of zero gravity. He would put small animals in the nose cone of these rockets and put a camera at the apex. And, then, as the rocket got up to the top of its apogee and started to come down, before

James Lovell
Former NASA Astronaut and President, Lovell Communications

Captain Lovell was selected as an astronaut by NASA in September 1962. He served as Command Module pilot and navigator on the epic six-day journey of Apollo 8, man's maiden voyage to the Moon. On that flight, Lovell and fellow crewmen, Frank Borman and William A. Anders, became the first humans to leave the Earth's gravitational influence. He was spacecraft commander of the Apollo 13 flight, 11-17 April 1970, and became the first man to journey twice to the Moon. Captain Lovell held the record for time in space with a total of 715 hours and 5 minutes until surpassed by the Skylab flights.

it reached terminal velocity, he would photograph their reaction to see how they would react in zero gravity.

Now, my story takes place out there at White Sands. One beautiful, blue day, out on the launch pad, is this dilapidated old V-2 rocket—gaseous oxygen just streaming out from the vent. Inside the nose cone there are two mice strapped tightly to their couches. This one mouse looks a little worried. His tail is twitching back and forth, and perspiration is coming out on his whiskers. He looks at his companion and says, "You know, I'm getting scared. The rocket could blow up! The parachute could fail to open! A mouse could get killed doing this kind of work!" And his companion, who had made about three flights before, said, "It beats hell out of cancer research!" So, in this particular case, this mouse figured that risk was the lesser of two evils.

Now, I'm going to tell you another personal story about this idea of risk. I'll go forward quite a bit to Gemini 7, Frank Borman and I are on a two-week mission—the purpose was to find out if man could live in space for two weeks, the maximum time to go to the Moon. And here is a case where, because of the newness of the situation, that risk was way overblown. The Gemini spacecraft proved to be a fairly decent vehicle; Gemini 3, 4, 5, and 6 were pretty good. But in those days, NASA and the doctors and the hierarchy—management—put the astronauts in the spacecraft and got them to keep their suits on all the time to fly these missions. For the first couple of missions—three and a half hours or even one day—that's fine. But, as time went on, those suits got to be more uncomfortable all the time, you know, oxygen flowing through the body, drying up the body pretty badly.

So by the time Gemini 7 came around, a two-week mission, we were determined that we were going to get out of our suits. We had a special suit but it was still bulky and uncomfortable. So we took off—and the first thing we then wanted to do was get out of the suits. We found out that the spacecraft's integrity was there. Nothing was leaking. Everything was fine. Management said, "No. No—stay in those suits." We said, "But everything is going fine here." Finally, out of desperation, I had unzipped my suit and I had snuck out of it (or almost), and I was out of my suit in everything but name. Poor Borman was still in his suit, and I could see he was getting more tired and difficult. And, finally, after about three and a half or four days, we finally got permission to get out of the suits.

So, here's a case where the risk was overkill. I mean, we knew the spacecraft was good. We knew the best way to fly was in our underwear, not the suits. And now, of course, as you and I see on TV, on the shuttle flights they're in shorts and T-shirts, so that's the way that goes.

And then the third little story I want to tell you about risk is one that you all know, but I think it's a classic. And it was the Apollo 8 flight. Apollo 8 was going to be an Earth orbital mission—around the Earth to test the Lunar Module and Command Module before we'd ever commit those two vehicles to go to the Moon. And as you know, two things happened in the summer of '68. Number one, Grumman Aircraft finally bit the bullet and said, "Hey, we're not going to get this Lunar Module ready before 1969." And then again, we had intelligence information that the Soviets were going to put a man around the Moon, a circumnavigation flight around the Moon, before the end of 1968—in fact, in the late fall of 1968.

And, as a matter of fact, we know now, talking to them and with everything in the open, that they were very serious about it. Their N-1 big lunar rocket was a failure, but their Proton and Zond probably could have done the job. And, so, I think in the fall or summer of '68, they sent Zond 5 around the Moon with small animals. I think the reentry was so steep that the animals died, but it was a test that they were doing to see if they could put two cosmonauts around the Moon. They sent another spacecraft—Zond 6 Proton went around the Moon again. And while that flight was not a complete success, it had the possibilities of success.

And here's where the change took place. In the Soviet Union, the hierarchy— the management—was arguing: "Is the risk worth the reward of beating the Americans at least to get two guys around the Moon, or should we send another unmanned or animal-bound flight around the Moon before we commit to the people?" Leonov and Makarov, the two cosmonauts, were all set to go. They were arguing: "Let's go." Other people said no.

SO HERE WAS A CASE WHERE WE ANALYZED THE RISK AND WE THOUGHT

THAT THE REWARD—THE ACHIEVEMENT AND THE ABILITY TO CONTINUE

THE APOLLO PROGRAM FOR LANDING—WAS WELL WORTH IT.

And while they were hesitating—while they were vacillating back and forth—a bold decision was made in this country, in the fact that the Lunar Module was not ready, but Apollo 7 showed that the Command Service Module could last for 11 days. And so, the decision was made to send Apollo 8 around the Moon and to look for landing sites and things like that.

So here was a case where we analyzed the risk and we thought that the reward—the achievement and the ability to continue the Apollo program for landing—was well worth it.

So, let me first state that everything in life involves a degree of risk— and I think I've mentioned that before—from the moment we are born until we die. And the risk can involve physical, financial, or emotional factors. You know, the Hollywood stuntman has to weigh the reward for his efforts to the risk he faces. The investor faces a risk of financial gain or loss. And, certainly, when we get married, the emotional risk is there for a happy marriage or a quick divorce. Therefore, when we have control of our destinies, such as an active space program, we must analyze the reward we achieve for the risk involved and the action we must take to minimize that risk.

In the space program—at least the one I knew—we approached the risk factor in many ways. First, the contractors, of course, set standards for maximum reliability—99 percent, if possible. And they used the concept of redundancy, you know—one of this or two of this or three of that. In case one failed, we had backups. Every effort was made to simplify space system design. One example: In the Lunar Module propulsion system, pressure-fed fuel systems were used instead of the more complicated pump systems. We incorporated escape systems. Our design of the trajectory to the Moon—the first part—would be a free-return course. That meant if the spacecraft's main engines failed in its inflight test, the spacecraft would be on a course that would take it to the Moon—and the Moon's gravity would aim it back towards the Earth. And by using only the spacecraft's attitude rockets, it could safely land back here on Earth. Thus, an added safety factor was given to the mission.

Of course, the intense systems training by the Mission Control team and astronauts was essential—including an analysis of possible failure modes and training to recover from them. Now, this training pointed out the limits on efforts to reduce risk in an Apollo mission. We only trained for single-point failures. Had we tried to train and develop recovering techniques for all possible combinations of failures—well, we'd still be at Cape Canaveral waiting for the first takeoff. And therein lies the problem between risk and reward.

I guess the best way to visualize this, at least from my point of view, is to picture a simple X/Y graph—a plot. Let's say that at the top of the Y, the ordinate at the top, is a factor up there saying "maximum risk." And then as we gradually go down the Y ordinate, the risk decreases all the way down until we get down to the juncture of the Y and the X graph—and there, theoretically, is zero risk. On the X axis, we put all those factors that we might be able to make in terms of cost—those factors that we can put into a spacecraft that would reduce risk—high reliability, redundancy, extra safety equipment that would cover any failures, true training, et cetera.

I kind of think that as we plot the graph going down, that the risk would decrease very rapidly until we got to some point where it would start to flatten out and keep parallel, never getting down to zero risk. As a matter of fact, I also think that had we continued to go out, adding additional redundancies, adding other equipment to handle other failures that might occur, and giving the crews more intense training, more procedures that they had to follow in case there are certain things that go wrong, that the risk factor would actually start to go back up again. Therefore, there's got to be a point whereby we can develop a system that we minimize the risk but without going overboard, because eventually you'll compromise the spacecraft's ability to complete its assigned mission.

Now, I think we did a fairly decent job in weighing the acceptable risks with effort to reduce risk in the Apollo program. The first six Apollo missions proved that. On Apollo 11, Mission Control quickly resolved the landing radar problem. The brilliant analysis by John Aaron saved the Apollo 12 mission after a lightning strike on takeoff. And so, by Apollo 13, Mission Control people and spacecraft crews were confident that they could handle any situation. There was, however, a wild card in our assumptions, and it surfaced on 13. Now Apollo 13 was the third lunar landing mission and strictly, I think, the first scientific flight. It was targeted to land in the hills surrounding a crater called Fra Mauro. The scientists thought the lunar material there would be different from that in Apollo 11 or 12 and, of course, we thought the surface there would tell us about the interior of the Moon.

The launch occurred on April 11, 1970, at 13:13 Central Standard Time. Perhaps the spacecraft number and the time was sort of a premonition of the events to come. During launch phase, our first crisis occurred. The center engine on the booster's second stage shut down two minutes early due to a "pogo effect" or extreme oscillation on its structure. Now, this pogo effect was noticed in one of the booster's unmanned flights.

To reduce the risk in this area, an engine shut-down device was added to prevent the engine from going divergent and disintegrating. In addition, the booster was slightly overbuilt to allow a one-engine failure. Here was an example of added safety features to reduce the risk of a flight. Our initial trajectory to the Moon was that free-return course that I mentioned. But at 30 hours after launch, we changed our flight path to what we call the hybrid course. Now this was necessary to provide the proper visibility for a safe landing in Fra Mauro. And here is where we traded the reduced risk of a safe return home for the guarantee of a good visibility. Should our spacecraft engine fail now, our closest point of approach to the Earth on our return would be about 2,500 miles out. Much too far out for a safe capture by the Earth's atmosphere.

We didn't worry about it. Fourth flight—second time to the Moon—and I was getting complacent.

The explosion took place two days and 200,000 miles from Earth, resulting in the loss of all the oxygen, electrical power, and propulsion of the Command Service Module. At this point, the flight of Apollo 13 changed from another thrilling space adventure to a classic case of crisis management. It was here, too, that other factors came into play to reduce the risk involved in spaceflight.

These are the attributes, or human characteristics, of a well-trained Mission Control team: good leadership—not just at the top—but throughout the organization, leadership that develops teamwork among all those involved, including contractors; use of initiative to find solutions to problems never contemplated or trained for; the ability to focus and persevere to find the right solution for each crisis; and, of course, a team that was well motivated to get the job done. Now, these are the ingredients that turned Apollo 13 from an almost certain disaster into a successful recovery. Mission Control and the flight crew worked together to configure the Lunar Module into a lifeboat. The crew successfully transferred the controls to the Lunar Module just as the Command Module died. Procedures were developed to use the Lunar Module landing engine to put the spacecraft back on a free-returning course.

Let me digress a little bit on this. There is something that I had learned in the space program, based on what I am about to say, that I took with me from the public sector into the private sector: Always expect the unexpected. When everything is going right—when everything looks rosy, when nothing is wrong—it's always nice to look ahead to see if there are symptoms coming down that maybe are pending for a possible crisis.

When I started to maneuver—now remember, I have two spacecraft mated together and I'm controlling from the Lunar Module, and remember, also, that I spent many, many hours in simulators learning how to fly a Lunar Module. But when I put an input in to make a certain change of attitude, the spacecraft didn't respond that way. I couldn't figure out why. If I wanted to go down, it went up. If I went left, it went right. I mean, after all these hours! Well, then it dawned on me. I had a 60,000 pound dead mass attached to the Lunar Module , the Command Service Module, which, of course, we needed to get back into the atmosphere. The Lunar Module had never been designed to be maneuvered with the Command Service Module attached. We had to quickly figure out how an input would give me the right output to get to the proper attitude to make that burn to get back on the free-returning course.

Now, again, it was discovered that the crew was being poisoned by their own exhalations. The round canisters in the Lunar Module to remove the carbon dioxide were becoming saturated. In the dead Command Module there were plenty of unused, square canisters.

Using their initiative, the crew systems division thought up a way to use tape, plastic, cardboard, and an old sock to adapt a square canister to the Lunar Module. This removed the over-abundance of CO_2 in the Lunar Module and, of course, prevented the poisoning of the crew. And, so, there was another little incident in system design. Why we had square canisters in the Command Module and round canisters in the Lunar Module, I will never know to this day.

Throughout the return home, the risk of disaster decreased and the odds became more positive as each crisis was analyzed and a solution developed. When it became apparent that the spacecraft would miss the narrow return corridor for a safe landing, a procedure was used that was developed as a last ditch measure for Apollo 8. I was on that flight as a navigator, so I happened to know about it. Using the Earth's terminator as a guide, a seat-of-the-pants manual maneuver was accomplished to put the spacecraft back on proper course. Again, proper training, including an analysis of how to make course changes after experiencing navigational failure, saved the day.

If, in the development of the Apollo program, we carefully balanced the risk versus the reward of a lunar landing by incorporating such factors as extreme reliability, redundancy, simplification, and intense training to reduce the risk, then what happened on 13? Apollo 7 through 12 succeeded in doing their missions, and the problems they encountered were easily solved by Mission Control working with their crews.

The answer is human error. It's a virus that can be embedded in the best laid plans. Those of you familiar with the causes of aircraft accidents will understand that most accidents are caused by a series of events that overcome the pilot and/or the aircraft. Such was the case with Apollo 13. The first event occurred about eight years before Apollo 13 took off. NASA ordered all Apollo contractors to make their electrical systems compatible with the 65 volt DC power available at the Kennedy Space Center—even though the spacecraft were designed to fly with a 28-volt DC power system. That would simplify the testing at KSC. The contractors complied with this request with one exception. A thermostat, part of the heater system inside the oxygen tanks, was not exchanged for one that could handle the high voltage. The job of the thermostat was to protect the tank from overheating. When the temperature rose to about 80 degrees Fahrenheit, the contacts would open, shutting off the heater power. At 65-volts DC power, however, the contacts could be welded shut, thus bypassing this safety feature.

All tanks on Apollo 7 through 12 had this anomaly, but none experienced the sufficient heater operation during testing to damage their thermostats.

A second incident occurred during the oxygen tank manufacture. A tank, designated for Apollo 10, was dropped at the factory. It was retested for flight qualification, but, because of the lost time, it was reassigned to Apollo 13.

Several weeks before the launch of 13, the third incident took place. With the booster, the spacecraft all assembled on the launch pad, a countdown demonstration test was performed, making sure that all the components were ready for launch. The test was successful, but after the test, the ground crew could not remove the liquid oxygen from one of the spacecraft tanks. A review of the history of the tank revealed the damage incident at the factory. Studying the design of the tank indicated that, although the tank performed perfectly for

. . . IF YOU'RE GOING TO HAVE AN ACCIDENT ON THE WAY TO THE MOON,

OUR RESEARCH SHOWS BE SURE YOU HAVE IT 200,000 MILES OUT.

all inflight operations, the fall could have impaired the ground crew's ability to remove the oxygen after a ground test.

To replace the tank would slip the launch by a month, and so the decision was made to use the tank's heater system to remove the oxygen by boiling it off. The procedure was successful, but as the level of the liquid oxygen decreased, the temperature rose. At 80 degrees, the contacts of the thermostat started to open to shut off the power. The high voltage welded them shut, and the thermostat, instead of shutting off the power, became a conduit to keep the heater system on. We know now that the temperature rose to about 1,000 degrees Fahrenheit, severely damaging the heater system. The problem was not detected. When the tank was filled with liquid oxygen, it was a bomb ready to go off. It exploded two days later, 200,000 miles from Earth, when we turned on the heater system.

I might digress another little bit here because, in all this discussion of risk, there is a factor that's called fate, luck, or something like that. This was the third time we turned on the heater system; nothing happened the first two times. If something happened the first time we turned on the heater system and that explosion occurred, we would never have had enough electrical power to get all the way around the Moon and get back home again, as we had already put the velocity on to go to the Moon.

If it did not explode when it did, but waited until we turned on the heater system later, once we were in lunar orbit or when the Lunar Module was on the surface, we would never have had enough fuel in the Lunar Module to either get out of lunar orbit, or get enough to get back home again.

So, if you're going to have an accident on the way to the Moon, our research shows be sure you have it 200,000 miles out.

I asked Gene Kranz, who was the lead flight director, what lessons he learned from Apollo 13 that could be applied to the Mission Control team—and maybe all of NASA. Here are some of his comments: Develop the chemistry of a winner. The mind-set for success must be embedded in the values and culture of the organization. Be positive. Be optimistic. Do the right thing the first time.

A second comment that he made: articulate a common vision that focuses your energies on your objective; team focus to accomplish the mission, whether it is in crisis mode or whether it's the entire organization. This was outlined by the President just recently. We must focus our energies on accomplishing that mission.

The third thing he mentioned: teamwork provides the multidisciplinary capability to deal with complex and fast-moving problems. We can say many brains are better than one. Get the team together. Think up the solution. And, I kind of think, when I look back now on our Apollo program, that this was pretty common throughout our entire NASA organization. We had good leadership at Headquarters. Marshall did the booster. Goddard did the network. Johnson did the spacecraft and the crew training. Kennedy did the launch and the integration of the whole thing. So, we had a pretty good team.

Fourth, Kranz says: build momentum quickly. This allows rapid response to limit problem growth. I think what he means there is that a quick response will give an insight to head off future problems that might be the result of an original problem.

He also says: be flexible. Solutions often lie outside the box. The idea there was the carbon dioxide incident on 13.

And then he says also: don't get distracted, and don't let your team get distracted. For Apollo 13, on that particular flight, when I was waiting for the information to come up to re-energize to get the Command Module back in operation again, there was delay after delay, and I thought that they were going to set up more information to find out what went wrong and give us more things to do than just get the spacecraft ready. I didn't want the crew down there at Mission Control to be distracted. I needed those basic procedures to get the Command Module going again.

He then says: overwhelm the problem. Use every available asset. As soon as you have one, call in everybody who has any idea of what may be happening, almost like verbal popcorn, but then you can winnow out what is good and what is not good.

Finally, his idea is: keep the poise. Let your words and actions convince your team that you are controlling events. Good leadership. You saw the movie. Gene Kranz, like Ed Harris when the whole Control Center is talking about finding out what went wrong when they found out about the explosion, says, "Stop guessing. Stop guessing. Let's work the problem."

In our approach to accomplish the President's directive of revisiting the Moon and on to Mars, we must accept a certain amount of risk and realize that unforeseen events are always present. The strategy of spacecraft development and mission design is to minimize the risk without compromising the goal. Whenever you are involved in an operation that handles thousands of pounds of high explosives, reaches extreme velocities, operates in a vacuum environment under zero gravity, and then encounters tremendously high temperatures on return, you are, if I can borrow the title from one of Tom Cruise's old movies, in a "risky business." The people involved in that business and those who monitor, critique, and investigate the results, should recognize that fact.

To be completely risk-averse is never to take off.

We should be aware that sometime in the future, we will again hear those words: "Houston, we have a problem," and I hope we'll be prepared to meet the challenge. Or, if I can steal the words from Gene Kranz, "failure is not an option".

NASA Image/Renee Bouchard

Bold Endeavors:
Lessons from Polar and Space Exploration

I want to tell everybody here that it is a pleasure to be here today to talk to you about some of my research. The concept of risk is something with which we all are familiar. Every decision that we make from the most trivial to the most important is attended by some sort of evaluation and consideration of the costs and the benefits, and the likelihood of a successful outcome.

Expedition risk is of a different order. And humans are not particularly good at estimating risk. The research shows that we have a tendency to underestimate risk over which we have some control, and to overrate risk over which we have no control. That's why we take the risk of driving on the highways, where presently there are 1.5 fatalities per 100 million miles traveled—incidentally that's down from 5.5 fatalities per 100 million miles traveled in 1966. You were four times more likely to die in a traffic crash 30 years ago than you are now, and there are nearly twice the number of automobiles and vehicle miles traveled. We've done a lot to reduce risk in certain areas.

But why do nations and individuals explore? I have here just a partial list. Trade routes, looking for new resources, in some cases national prestige, and, of course, science. Individuals explore sometimes to satisfy a need for achievement, to do something special,

Jack Stuster
Behavioral Scientist and Vice President and Principal Scientist, Anacapa Sciences, Inc.

Jack Stuster's work for NASA has included a study of Space Shuttle refurbishing procedures and studies of conditions on Earth that are analogous to space missions, including an analysis of diaries maintained by the leaders and physicians at French remote duty stations in the Antarctic and on small islands in the South Indian Ocean. He has developed design and procedural recommendations to enhance the habitability of the International Space Station, future spacecraft, and planetary facilities. Stuster completed a study of Antarctic winter-over experiences, expeditions, and voyages of discovery, which are documented in his book *Bold Endeavors: Lessons From Polar and Space Exploration*, published in 1996 by the Naval Institute Press.

many times out of curiosity, including scientific curiosity, and I truly believe that some people explore because they need to accept risk. Life just isn't enough without taking some chances. However, taking calculated chances is far different than being rash.

Every bold endeavor that I've read about was accompanied by naysayers, people who predicted that the expedition would result in disaster. It's archetypal that Columbus had difficulty finding the financing for his planned expedition. It wasn't because people believed the world was flat. By 1492 all learned people knew that the world was a sphere. The circumference of the Earth had been calculated by the Greeks, and then again later, and accurately, 400 years B.C. or so, and again later, but the later estimate was off by a large factor.

Columbus believed that he would reach Japan after traveling about 3,200 miles west. He was right. He did make landfall 33 days after leaving Spain. But had he known that it was really 10,000 miles to Japan, and that a continent or

THE HISTORY OF THE HUMAN RACE IS A CONTINUAL STRUGGLE FROM DARK-
NESS TOWARD LIGHT." I THINK THAT'S BEAUTIFUL. "IT IS THEREFORE TO NO
PURPOSE TO DISCUSS THE USE OF KNOWLEDGE. MAN WANTS TO KNOW,
AND WHEN HE CEASES TO DO SO, HE IS NO LONGER MAN.

two interrupted his voyage, he might not have taken that risk. He did maintain two journals, one for his own use, and one for the crew that showed they were making far greater progress than they actually were—a way for him to minimize his personal risk on board.

There are many justifications for exploration. One of my favorites is from Fridtjof Nansen, a Norwegian explorer, that might seem appropriate in this age when people complain about spending money on space. I mean—the critics say we should spend it here—as if the money were actually taken into space and thrown out of the spacecraft. But Nansen, who was a scientist as well as an explorer, wrote that "people perhaps still exist who believe that it is of no importance to explore the unknown regions. This, of course, shows ignorance. The history of the human race is a continual struggle from darkness toward light." I think that's beautiful. "It is therefore to no purpose to discuss the use of knowledge. Man wants to know, and when he ceases to do so, he is no longer man." I think that says it all. And also, Nansen was an early supporter of women's suffrage, so please don't judge him by his 19th century usage of the term 'man.'

Roald Amundsen was a little more blunt in saying that "Little minds only have room for thoughts of bread and butter." But I will talk more about both

Nansen and Amundsen in a few minutes. There are many things I want to talk about that I'm sure I'm going to forget, so forgive me for that.

Robert Falcon Scott wrote, after his first expedition to Antarctica, about how ill-prepared they were. "Not a single article of the outfit had been tested, and amid the general ignorance that prevailed, the lack of system was painfully apparent in everything." Robert Falcon Scott gave great advice about things, but he didn't really take his own advice. In his final hours, having reached the South Pole in 1912, only to find that Roald Amundsen had been there 30 days earlier, and on the trip back, laying in his tent with comrades who had perished beside him, he wrote in his journal that "We took risks, we knew that we took them. Things came out against us, and therefore, we have no cause for complaint."

Scott was unlucky also. They perished only 8 miles from the supply depot that had been prepared for them. They just couldn't get to it in the storm—1912 had been an unusually stormy year in Antarctica. Under other conditions, they might have made it to the depot and come home to write an account of their expedition.

Apsley Cherry-Garrard, who was also a member of Scott's expeditions, wrote that "the members of this expedition believed that it was worthwhile to discover new land and new life, to reach the South Pole of the Earth, to make elaborate meteorological and magnetic observations and so forth. They were prepared to suffer great hardships, and some of them died for their beliefs."

They should have been more prepared. Others were. Scott used Manchurian ponies, which didn't really cut it in the snow, nor had they ever tested the tractors they took to Antarctica. There was a certain hubris involved. Amundsen used dog sleds. The British would not use dogs or skis. It wasn't British. They were going to slog it out.

Most of my work has involved the risks associated with the psychological, behavioral, and human aspects of isolation and confinement. I use the following analogy to help people get a handle on what it would really be like to be on an expedition to Mars. Imagine living in a motor home with five other people for three years. You're driving around the country, and you really can't get out for about a year, and then, when you go outside, it's for very brief periods, and you have to wear spacesuits, and you come back, and then you spend another year or so driving around with those same five people. You've already heard every story that they've ever told. The days blend one into another. The condition becomes mind-numbing, and the tiniest, tiniest things get on your nerves. It is characteristic of all conditions of isolation and confinement that trivial issues are exaggerated way out of proportion. Everyone who I've interviewed about this talks about how they would have an incredible argument at an Antarctic research station over a fax transmission or something, and blow up, and then an hour later wonder: "What the heck happened? What was that all about?" It is a universal occurrence.

One of the other universals of isolation and confinement is the strange relationships that occur with your Mission Control, with your headquarters, wherever it is located—in Antarctica, it might be Port Hueneme, or it might

23

be the Johnson Space Center or elsewhere. But the remote crew always gets the impression that "They really don't understand the conditions under which we're operating. We're trying to get a job done here and they're not responding fast enough." Or, "They're giving us too much to do." It always happens. And, you know, I used to think that it was just endemic to isolation and confinement, but I think it's a structural condition. Even the field offices of a corporation, a small one or a large one, or perhaps the research centers of a major government agency might feel these same sorts of tension. It is just a natural phenomenon that occurs. If you're prepared for it, you can somehow reduce the risk.

Anyway, an expedition to Mars would be a lot like this metaphor that I've described for you. The first research that I conducted for NASA was conducted for the Ames Research Center. In 1982 they took a chance on this anthropologist who was working in the field of human factors to study conditions on Earth that are analogous to what we expected for future space crews. I studied conditions such as offshore oil platforms, commercial research vessels, fishing vessels, fleet ballistic missile submarines, saturation divers, and so forth, and came up with 100 or so design recommendations. It's my understanding that a couple of them actually made it to the final design of the International Space Station, for which I'm grateful. I would like to know which ones they are. Personal sleeping quarters I don't think has made it, and that was one of the most important recommendations.

More recently, I've conducted research through the Johnson Space Center concerning longer-duration missions, one year to three years. The only analogues available for such a long mission are previous expeditions. And, of course, I included our experience with Skylab, and there is much of relevance from Skylab.

NASA has a tradition of trying to learn from the past, and in many cases is successful. However, I remember reading in one of the industry publications that: "One of the great lessons from the NASA experience on board *Mir* was that you really shouldn't hard-schedule everything. You should have this task list that you put things on. And then the crew can go and take from that task list as necessary. Isn't that a wonderful thing?" I thought: My gosh, that was the principle behavioral finding from Skylab. Didn't anybody read those wonderful lessons learned reports from Skylab?

So, I wrote a letter to the editor, and I probably angered a whole lot of people in doing so, but there is a lot that we can learn from the past, including our own more recent past.

I've found that expeditions, and polar winter-over experiences in particular, resemble in many ways what we can reasonably expect for future space crews. Chronologically, the earliest of the expeditions that I studied was Columbus's first voyage of discovery. And although it was only 33 days out to the New World and seven months total, there really is a lot to learn from that experience. For example, he had strong-willed subordinates who questioned his authority regularly. One of them [Pinzon, commander of the *Pinta*] left the expedition in search of gold to the north, leaving the two principal vessels.

And it's probably not well known that on Christmas Day, 1492, the *Santa Maria* went ashore and was broken up. The reason was the crew had partied the night before, celebrating Christmas Eve, and left the watch to a cabin boy who didn't know what to do when the ship slipped its anchor. No one was killed during the process, but it left Columbus with only one hull.

Columbus believed in triple redundancy long before it was a NASA policy, and he probably would not have left Europe with fewer than three hulls, and certainly would not have returned. Oddly, in one of those incredible coincidences that occurs that I've read about in the history of exploration, Pinzon rounds the bend of this little island—this tiny island where the crew was trying to decide what to do. Would they be able to rebuild and make a small craft out of the remnants of the *Santa Maria*? And then Pinzon shows up. They were able to return home, but in the two smallest of the three craft.

Redundancy is an important method for reducing risk and increasing reliability. There are other methods: overbuilding—you build the valve to withstand 150 percent of what you expect it to withstand; graceful degradation, so that you have time to do something about it; and maintainability. When you have a human crew, you should really take advantage of the crew for maintainability.

One of my favorite explorers is the French explorer, Jules-Sébastien-César Dumont d'Urville. Early in his career, he was on the island of Milos when people approached him about a statue that was hidden in a cave. He saw it and wanted it for France, so they dragged it down to the ship, breaking off two arms in the process. It's what we know as the Venus de Milo. Later in his career, he commanded two expeditions to the Pacific and to Antarctica. He was one of the first to see the mainland of Antarctica, which he named Adelie Land for his wife, whom he rarely saw. He also named the linguistic groups of the Pacific with the names that we use today—Polynesian, Melanesian, Micronesian. He was an exceptional leader. At a time when expeditions—naval ships, in particular—were commanded autocratically, he was a kind and generous captain. He dressed as the crew did, which perplexed the British any time they met, because they didn't understand. They didn't believe he was truly the captain when he was wearing a straw hat and an open shirt. He was a realistic man.

On his second expedition, he was required to leave Marseilles carrying plants to the South Pacific. I don't know exactly what the plants were, but he had lots and lots of plants. At first, he objected to it because they were in pots and all over the ship, including in his cabin. And, after a week at sea, he wrote in his journal that this was a wonderful addition to an expedition and, if he had his way with things, every French ship that left port would be accompanied by plenty of foliage and greenery inside. I think that that's not too dissimilar from some of the comments that we've heard from space crews loving to spend time with the growing experiments on board.

The French had discovered early on something that was very painfully learned elsewhere, and that is, that there's often conflict among subgroups in an isolated and confined situation, and there were a lot of problems with the civilian scientists and the military crew. The scientists were outside of the command structure and it was always a problem, which led to the demise of some expeditions, or contributed to it, at least. So the French would take bright Naval officers and train them to be botanists or natural philosophers and artists.

It's particularly appropriate that we talk a little bit about the Lewis and Clark expedition in this year of the bicentennial. And there is much to learn, even though there are great differences. It was all outdoors, for one thing, and not in a confined environment, except when they were in winter quarters in Oregon where it was raining all the time. One of the things that we can learn from the Lewis and Clark expedition is to establish a spirit of the expedition. Thomas Jefferson named it the "Corps of Discovery"—a brilliant thing to do. I was very pleased in 1999 when I visited the Astronaut Office at Johnson Space Center and saw a sign that read," Expedition Corps." I asked, "What is this?" Andy Thomas responded, "Well, it's for the people who are planning to go to the International Space Station and beyond." I said, "It's a stroke of brilliance." You have people already using the mind-set that this is an expedition. It's going to be a long time—it's not a test flight, it's really an expedition. It's my understanding that Michael Foale is responsible for doing that. [Foale replied that astronaut Ken Bowersox (also in the audience) was responsible for the use of the term]. Well, it was a stroke of brilliance and should be congratulated. It's a wonderful idea. It helps people get in the mind-set for an expedition.

There were 40 explorers with Lewis and Clark. By the way, only one member of the expedition perished in the entire three years, and he died of a burst appendix, we believe, based on a description of the incident. Any one of you who ever had acute appendicitis would probably agree with me that you'd want to have that out before you go. Now, the physicians tell me that that's not necessary, but, from my experience, I wouldn't want to have that condition a long way from home. The Lewis and Clark Expedition was 28 months long, about the same as an expedition to Mars might be.

Lewis and Clark and their company met many native peoples along the way. That probably won't happen on a mission to Mars, although some people are hoping for it, I'm sure. But one thing that they did was to describe everything in their journals. Captain Clark and Captain Lewis were meticulous journal keepers.

I thought it might be interesting to find out what exactly they were doing on the 27th of September 1804—200 years ago today. I was amazed. It was the most pivotal period of the Lewis and Clark expedition. Two days ago, they were on the Missouri River, and they reached a tributary near what is now Pierre, South Dakota. They had finally encountered the Teton Sioux, who they had heard were going to be hostile to them. Indeed, it was a three-day period of intense hostility. They had learned through interpreters—through other Native Americans—that the Sioux intended to prevent them from going any farther and to steal all their stuff. The two preceding days were just incredibly tense.

On the 27th, they were trying to leave the village, and the little boat that was taking them out to the larger keel boat had lost its anchor and was having trouble maintaining its position. The little boat came out and parted the remaining cable, and there was a lot of hollering to get the people to their oars and so forth, and that alarmed Black Buffalo on shore, so he called all 200 of his warriors out to the shore. Lewis and Clark believed for sure that this was going to be the showdown. They went to stations—Clark went to the bow and manned the swivel gun, a little two-inch cannon loaded with shot. They had something like 20 men with blunderbusses loaded with shot trained on the main body of the group. They had a technological edge here. They would have wiped out 40 or 60 of the Teton Sioux, but there's another 200 of them in arrow shot, and they could keep an arrow in flight at all times, and it's a long time to reload the weapons on board the keel boat.

There was this standoff for we don't know how long, but it appears to be quite a while, with Clark in the bow shouting, the interpreter, who really didn't speak Teton Sioux, trying to convey to Black Buffalo to control his people

ONE OF THE THINGS THAT WE CAN LEARN FROM THE LEWIS AND CLARK EXPEDITION IS TO ESTABLISH A SPIRIT OF THE EXPEDITION . . . IT HELPS PEOPLE GET IN THE MINDSET FOR AN EXPEDITION.

because there were warriors who were coming into the water, who were grabbing hold of the mast of the little boat to keep it ashore. They thought for sure that this was the incident that they had been fearing. What Clark didn't realize was that his people obeyed him because it was a military organization. The Teton Sioux were only recently a tribal organization. It was a group of bands that came together when the resources permitted. Black Buffalo's control over the 200 or so was based on his charisma—only a quarter of them were related to him and had some obligation to obey him. But Clark took a risk that if he held his ground and didn't fire, it would be resolved peacefully. And the decision paid off. Finally, Black Buffalo pulled on the arm of one of the guys and apparently told him to back away, and the Corps of Discovery was permitted to go.

Of course, the Sioux dogged them all along the way, trying to get them to come ashore or to take them on board, which Lewis and Clark didn't do. I'd just like to read a sentence or two from the journal entry for this day 200 years ago. "We were on our Guard all night. The misfortune of the loss of our Anchor obliged us to lie under a falling bank, much exposed to the accomplishment of their hostile intentions . . . Our Bowman, who could speak Maha, informed us in the night that the Maha prisoners informed him we were to be stopped. We showed as little signs of this Knowledge of their intentions as possible. All prepared on board for anything that might happen. We kept a Strong guard all night, no Sleep. Captain Clark, 27 September 1804." Just south of the Mandan villages is where this all occurred 200 years ago today.

The lessons applicable to the future? The importance of good leadership. Previous studies found that good leadership is actually more important than good habitability. Plan everything. Have a sense of cooperation and perseverance. To the extent possible, live off the land. Now, you won't be able to hunt buffalo on Mars, but you will be able to use the resources on Mars in the same manner to extend your reach. And, of course, develop a spirit of the expedition, symbolized by the Corps of Discovery.

Another expedition that everyone knows about is the voyage of the *Beagle*. It was really a British surveying expedition, the purpose of which was to chart the coastline of South America. Captain Robert Fitzroy was—I can't think of a polite word to use—a very stern and narrow-minded person. He at first didn't want the volunteer naturalist, Charles Darwin, on board, because he didn't like the look of his nose. And then later, off of the coast of Argentina, Darwin had an argument with Fitzroy and almost abandoned ship, because Fitzroy thought that slavery was a noble institution and had a lot going for it and Darwin thought it was disgusting. And, so, at their next port, Darwin spent several weeks on shore until he cooled off.

Darwin wrote in his journal about the crowded conditions on board a research vessel. So many chronometers and so many people packed into small space. It was a very difficult journey for him. Darwin, after this five-year voyage and returning to England, lived to be a very old man. But he never again set foot on a boat, never again left England.

One of the most relevant expeditions is the Belgian Antarctic expedition of 1898–1899. It's relevant not just because it was the first expedition to winter over in Antarctica, the first expedition to really have science as its true objective in Antarctica, but because it was a multinational crew, cosmopolitan, and, in this regard, truly modern. It included Norwegians, Romanians, and, of course, Belgians. They had the very best of all French food, and one American, Frederick Cook, the ship's physician.

What happened on board the *Belgica* is well-documented. The crew gradually slipped into a malaise that was paralyzing to some of them. One man died because of what Cook thought was the effects of the isolation and confinement. One man developed a temporary deafness. Another man developed a temporary blindness. One man, each night, would find a place below deck where he could hide and sleep, because he thought people were going to kill him. Roald Amundsen served his apprenticeship as an explorer as mate on the *Belgica*, and later wrote, "Insanity and disease stalked the decks of the *Belgica* that winter." He credited Frederick Cook with saving the expedition from certain psychological collapse.

Cook saw what was happening, and he thought that there was this heavy psychological component, but he also thought something was missing from their diet. This was before vitamins had been discovered, but he figured there was something missing. He tried to get the men to eat fresh penguin meat, but it tasted too fishy for many of the men. So, for those who were the most afflicted by this malaise, he would have them stand with nothing on except an overcoat exposing their naked skin to the glow of the ship's stove. He called it the baking treatment. They'd stand there for as long as they could each day, taking turns doing this. Whether it had some effect on them, or maybe it was a placebo effect, it did have the effect of helping the crew get through this very difficult period. Cook also thought that exercise would help, so he required the crew to take walks on the ice, but this devolved into a circular path around the ship that became known as the "madhouse promenade."

It was a dismal time, and it appeared when the spring came that they were not going to be able to release themselves from Antarctic's icy embrace. They worked very hard with ice saws and explosives and finally did break free, because they knew that they couldn't survive another year.

This is not to say that people haven't survived isolation and confinement before; many have. There were often several hundred whaling ships locked in the ice at any given time in the north during the 19th century. It is well known that during the height of the Cold War, there were 10,000 American submariners, at any given moment, at sea, in isolation and confinement.

Regarding the Australasian-Antarctic Expedition and Douglas Mawson, I formerly neglected the Australian contribution to exploration until my dear friend, Desmond Lugg, showed me that it was just a characteristic American narrow-mindedness to focus on certain things and disregard the rest. I rectified that situation by reading as much as I could about this expedition and about Mawson. There is a tremendous wealth of information that we can extrapolate from Mawson's experience. For one, personnel selection is important, and, for another, weather influences everything. It'll interrupt your plans. It will break equipment and keep you from doing things that you want to do. If you don't think that's relevant to the future, ask Michael Foale, who had on several occasions to retreat to the hardened portion of the International Space Station when there were solar events, solar weather. Also, on Mars, there will be similar solar events and solar particle events and also dust storms. Dust storms on the planet Mars can envelope the entire planet, and that would affect an expedition.

Roald Amundsen was the most successful of all explorers; he always made it to his destination. First to the Northwest Passage. First to the South Pole. In 1923 he was on two Dornier flying boats to fly over the North Pole. One of them developed problems and had to land. It crash-landed. The other one landed. They spent two weeks on the ice, leveling with wooden spoons an airfield for them to take off. Amundsen structured every moment of every day. The hours of work, the hours of eating, the hours of sleep, the hours for talking, for smoking, everything. He was in charge, and he made himself known to be in charge and organized everything. When they returned to Norway two weeks later, of course everyone thought he had died in the ice, and it was a wonderful welcome. Amundsen later perished in the North while looking for Umberto Nobile, a guy who he devoted his biography to criticizing. I work in the field of human factors, and I'm grateful

29

to Roald Amundsen for his wonderful statement, "The human factor is three-quarters of any expedition."

Ernest Shackleton is probably the best known of all the explorers. There are movies about him, books about him, and seminars at corporations to impart the style of leadership that he had developed. His recruiting ad from a London newspaper read, "Men wanted for hazardous journey. Small wages, bitter cold, long months of complete darkness, constant danger, safe return doubtful. Honor and recognition in case of success." Now, this might have been a personnel selection measure on his part, because I truly believe he thought he was going to return, but he wanted to make sure that everyone who embarked with him would be aware of the risks.

Shackleton had very clever ways of selecting people not so much on their technical expertise, but on how well they got along with their colleagues. He would ask them impertinent questions, and if they responded defensively, that might not be the kind of person that you really want in your tent eight months into a bad situation. But if they were humorous about it or philosophical about it, the person might be okay. Although Shackleton never made it to any of his destinations, he never lost a man. On the British trans-Antarctic expedition, the *Endurance* was locked in the ice, and [the] crew spent months on board, and then several months in a camp next to the ship as it was sinking. Then they moved to a camp that was on an ice floe that was as large as they could see, but, gradually, as the winter ended, the ice floe was breaking up around them. It was a mile across. Then it was several hundred yards across. Then it was 100 yards across. They had been practicing their egress to the boats. They had saved lots of equipment and three cutters from the ship. They had everything in the boats and they had practiced many, many times to escape the floe. It started to break up beneath them. It actually broke up right in the middle of the camp. Shackleton dramatically rescued one of his crew members from the ocean, pulling him onto the ice, and they departed. Then, they spent a week in these open boats in the worst sailing conditions on the planet, before they made it to a tiny rock called Elephant Island, where they made it ashore.

Shackleton knew that they could not survive there very long, so he selected five men to accompany him on the most arduous and dangerous open-boat voyage probably ever undertaken, to get to a whaling station on South Georgia Island. He took some of the people with him because he needed their skills, but he took some of the five people with him because he didn't want to leave them there. They were the malcontents that might have made things really bad for the folks who were going to be confined to the huts they made from the overturned cutters on Elephant Island. He eventually made it to safety. They made five rescue attempts, finally getting to Elephant Island with a borrowed tug from the country of Chile. It is a wonderful story.

I want to talk just for a moment about Richard Byrd, because he's American and one of the few of the American polar explorers that I consider relevant. On his 1934 expedition, Byrd built Advance Base, a 9 by 13 foot hut that was transported 100 miles from Little America and buried in the snow. It was going to be his experiment in isolation and confinement. Originally, he intended to have two people live there, but wrote later that he didn't want to subject anyone else to the risk. He considered the primary source of risk to be the psychological risk of being alone in complete darkness. Well, he really shouldn't have done this, because he almost killed himself three different ways. He fell and injured his shoulder even before the party that had delivered him had departed. He was continuously poisoning himself from the exhaust from the gasoline generator and from the fumes from a poorly vented stove. He almost froze to death when he locked himself out of the cabin in a storm—that was poor human factors preparation, the latch on the door.

But the crew at Little America knew that something was wrong several weeks into this experiment when his Morse code transmissions were the equivalent of slurred. They mounted three different rescue missions before they got to him, and he was in terrible shape. He survived to write one of the most eloquent accounts of life in isolation and confinement at its worst in the book *Alone*, in 1938. "Time was no longer like a river running, but a deep still pool," he wrote. He also said that "a man who lives alone lives the life of a wolf." That is, his manners left him, which is something that happens in isolation and confinement

The Norwegian Polar Expedition is one of my favorites and the expedition from which we can derive the most benefit. Fridtjof Nansen would have had a wonderful career in modern times, either as a rock musician or an actor. But he was a scientist. He was one of the founders of the modern theory of neurology. He was one of the popularizers of skiing as a sport. He had skied across Norway from Bergen to Oslo. Skiing was not a sport at the time, it was something rural people did to get around.

It is difficult for us to appreciate what the world was like during the closing years of the 19th century. We take for granted a communications network and travel abilities that allow us to reach anywhere in the world. But in 1893, there were still many unknown regions and many unanswered questions of the natural world, and the most compelling was, "what is at the North Pole?" Is it land? Is it ice? Is it open ocean? There were fanciful predictions. And many people had perished trying to find out.

Nansen had a plan. There was some evidence that the polar ice pack moved across the top of the world from east to west. So he thought: if a ship were built properly, it could be locked in the ice on purpose, and then you could allow nature

31

to carry you across the top of the world. He had a plan for a ship which he called the *Fram*. "Fram" means "onward" in Norwegian, and it was his personal motto. He approached the Norwegian government with this plan and received a grant. He had to go back, not unlike modern expeditions, because of cost overruns for building in an additional margin of safety.

During a time when crews were separated—with the "men," or crew, sleeping before the mast in the forecastle, and the officers and scientists in the main cabin—Nansen designed the *Fram* so that all staterooms opened onto the saloon, or the main area, a perhaps characteristically egalitarian, Norwegian approach. It was a very stratified society, but he did this to encourage comradeship and facilitate habitability. Nansen tested everything beforehand. There were spinoffs from his expedition. Polar travelers still use the Nansen Cooker, because it extracts the last calorie of energy from fuel.

The Norwegian Polar Expedition provided a model for all future explorers. The *Fram* sailed up the coast of Norway, across Siberia, and at a point closer to Alaska than Norway, headed into the pack ice on purpose. The ship was built with a rounded bottom and a recessed keel. Every fitting could be removed so ice could not get a purchase on this ship. When the ice encroached, and the pressures increased on the hull, the ship rose up out of the ice and remained cradled in that manner as she drifted across the top of the world. The theory was proved, and when it appeared they would get no farther north, Nansen selected one man, Hjalmar Johansen, to accompany him on a dash to the pole.

After many weeks, they found that they were only making a mile a day. So, at the closest that anyone had reached to the North Pole at that time, they turned back. They had no hope of regaining the *Fram*. They made it to Franz Josef Land where they were caught by an early winter.

Nansen knew that the secret was to keep people busy with meaningful work, and, of course, to be especially careful about the food. Norwegians are not afraid of the cold. They say there is no such thing as bad weather, only bad clothing. And he also knew that it was important to keep people entertained. The crew looked for every opportunity to celebrate. After awhile, they actually went into their almanac to find other countries' holidays to celebrate. Special celebrations break the monotony and help motivate a crew.

Nansen and Johansen built a 6 by 10 foot hut out of stones and walrus hides. Their entire world was illuminated during that Arctic winter by the pale glow of a blubber lamp. They had nothing to do. They slept sometimes as many as 20 hours out of the 24, in the same sleeping bag, because it was the most efficient way to conserve heat. But they never resorted at any time during their nine months to any sort of conflict or harsh words. This was the first thing that the press asked them when they got back. How did you survive?

They burst from their hut in the spring and performed every task that was required of them expertly, despite the mind-numbing sameness of the nine months that they had endured in isolation and confinement. They couldn't clean themselves. They had no towels. They didn't have a change of clothes. They would take their knives to scrape the soot that came from the blubber lamp that heated their food and illuminated their hut. They would scrape the blubber off and back into their lamps, recycling the fuel. It was incredible. Their dreams were filled with clean clothes and Turkish baths.

Nansen and Johansen came upon a British expedition within a month after leaving their hut, and they stayed there for another month or so until that expedition's relief ship came. The day that they stepped foot on Norwegian soil, the *Fram* broke loose from the Arctic pack ice on the other side of the world, then made its way back. The crew was united and sailed together around Norway and up Christiana fjord to what is now Oslo. They were greeted as if they had just returned from another planet. It's hard for us to imagine what it was like 110 years ago, but the similarities to the feelings that we would have are certainly there.

This artist, explorer, neurologist, oceanographer, champion skier, and founder of Norway was instrumental in the League of Nations. He received the Nobel Peace Prize for saving hundreds of thousands of lives from the Armenian situation, and also helped with a famine. The new Soviet Union after World War I wouldn't recognize the Red Cross. Nansen was respected throughout the area for his experiences, and organized a relief effort, when he found that there was a famine underway, while helping to repatriate prisoners of war. Presently, there are people in Eastern Europe who hold what is called a Nansen passport for displaced persons. His legacy is wonderful.

There is much to learn from the past that is applicable to the future. I have a lot to say about that, but I am out of time. The main themes to emerge from my research are: Certain problems are highly predictable, but they can be mitigated by taking the proper precautions. One of the most important findings is that humans can endure almost anything.

My work has focused on the behavioral and human factors issues, and I performed a content analysis of diaries that were maintained by the leaders and physicians at French remote duty stations on tiny islands in the south Indian Ocean and at the Dumont d'Urville station in Antarctica. Engineers have been asking the behavioral sciences for many years, "What's the most important behavioral issue? Is it privacy and personal space? Is it sleep? Is it group interaction? What is it?" Psychologists and others would say, "Well, group interaction." "Well, how much more important?" "We don't know."

I used content analysis to help answer the engineers' questions. The method is based on the assumption that the more someone writes about a topic, the more important it is to that person. I found that group interaction received almost twice the number of category assignments as any other category. The study resulted in the first rank ordering of behavioral issues based on quantitative data. I also found a decline in morale during the third quarter of an expedition; whether it is a 5-month mission or a 12-month mission, there is a drop, in effect. Initially, I thought, isn't this an interesting and useful discovery. Then I started to realize that it applies to almost everything. Think of a semester in college: you're only three-fourths of the way done and there is all that work yet to do, and

I've only got three weeks remaining. I think it applies to many situations in addition to isolation and confinement.

There are some specific lessons. One of them is to design for redundancy, as NASA does so well, and also for maintainability. There is no substitute for having Captain Lovell on board to take duct tape and fabricate a solution to a problem. One should expect casualties. Don't consider it out of the question. Also understand that weather will affect everything. The conditions will be different, but most of the problems that will confront future explorers will be the same problems that were confronted in the past. It won't be the gasoline-powered generator or the poorly vented stove that Byrd encountered, but some other similar situation.

We have embarked on a new age of discovery already, and there is much more in store for us—wonderful things.

One of my favorite quotes is from Arthur C. Clarke, who is one of the most prescient people on the planet. He invented the PDA for *2001: A Space Odyssey*. He invented the communications satellite, as we all know. His words inspire me. Every time there is a visible pass of the International Space Station over my house, I am out on my roof watching it. "Every age has its dreams, its symbols of romance. Past generations were moved by the graceful power of the great windjammers, by the distant whistle of locomotives pounding through the night, by the caravans leaving on the Golden Road to Samarkand, by quinqueremes of Nineveh from distant Ophir . . . Our grandchildren will likewise have their inspiration—among the equatorial stars. They will be able to look up at the night sky and watch the stately procession of the Ports of Earth—the strange new harbors where the ships of space make their planetfalls and their departures."

I could find lots and lots of quotes about taking risks. There are hardly any about not taking risks, which might be telling. Of course, we heard earlier about Admiral Zheng, whose armada of more than 300 ships in the early part of the 15th Century sailed from China all the way to Africa. The flagship of his armada was more than 300 feet long. Compare that to state of the art 1492 [European] naval technology. What would history have been like had the Emperor not had all the ships burned and made it a capital offense to build a ship with more than two masts? We might all be speaking Chinese now. I'm not sure. It's important, sometimes very important, to take risks, because the costs of not taking them can be greater than taking them.

I want to end on a more cheerful note. My favorite philosopher, Mark Twain, commented on more than the weather in San Francisco in the summertime. He also said "Twenty years from now you will be more disappointed by the things you didn't do than by the ones that you did do. So throw off the bowlines. Sail away from the safe harbor. Catch the trade winds in your sails. Explore. Dream. Discover."

Discussion

MILES O'BRIEN: We have a little bit of time left, and I just wanted to open it up. Raise your hand if you have a question.

QUESTION: The question I have for both Jim Lovell and Jack Stuster is: It seems to me one of the key differences between the explorations which you have studied so much and space exploration is something that you touched upon, that is, the relationship between the leader on site and the team at Mission Control. That, to me, seems to be a big difference between the polar missions where the leadership of one person on the ice meant everything and space missions which, ultimately, like it or not, will be second-guessed. What is the best way, as NASA plans missions of great length, to work that out so you don't end up with a Skylab mutiny-type situation?

JAMES LOVELL: Well, let me answer that question in this manner. In the early days of our space exploration, as many of our audience knows, the people who designed the work to be done on the spacecraft, sitting back at a desk and thinking of what to do, often had an overabundance of things to do until you got into the spacecraft. When you were actually working in zero gravity and you had the ability to adapt to that zero gravity, you were overburdened with things to do at the beginning. So, the people on the ground have to realize what the conditions are in the spacecraft to be able to accomplish the tasks that you give the crew. In the early days, this was a lot of times not thought about until the crews sort of rebelled and went back to the controllers or mission planners and said, "Look. Here's what we can do, and here's how we have to stretch out the agenda."

JACK STUSTER: You're right. The early explorers, of course, had no way of communicating with their base of operations. And, even when it became possible, some didn't take advantage. For example, Shackleton could have had a [radio] transmitter on board the *Endurance*—he chose not to, because he didn't want to have that connection. And I've discussed this issue with Claude Bachelard of the French polar program, and if he had his way, he wouldn't have much communication at all with home, only the most necessary things, because of the potential for problems. In that list that I mentioned very briefly of the behavioral issues, number one is group interactions. Number two was outside communications. And most of it had a negative valence to it. So communications is definitely an issue, but NASA is doing an awful lot in that regard.

The Life Sciences folks at NASA Headquarters have sponsored a great deal of research just on this very issue of communications between on-orbit crew and Mission Control. I watch every morning NASA TV, and the relationship that Mission Control folks have with the International Space Station crew is wonderful. It just seems terrific, and it goes both ways. The crew learns how to deal with Mission Control, and Mission Control learns to be sensitive to the special issues of the crew as well.

MILES O'BRIEN: Of course, on Mars you'd have a 40-minute roundtrip for communication. That would probably complicate things a little bit. It would be more like e-mail.

QUESTION: I just recently downloaded the Saturn I user manual that the Skylab guys referred to, and it says the specification for reliability on the Saturn 1B system was only 0.88. Now, with the Shuttle and how we've gone towards the all nines or five nines (.99999), where do we look at the boundary of reliability, which you were talking about in your discussion?

JAMES LOVELL: Well, it's tough to answer that. I think a general assumption that whenever we design any component, whether it's a booster, a spacecraft, or a segment of something, we try to get the greatest reliability. We try to man-rate the system so that we can have reliability on the system that we're going to use. Now, I don't exactly know what all the percentages are of the various items that we've used over the years. But I would assume that in our present operation, and in the future work on some of the new vehicles that we'll design, that is one of the greatest concerns and greatest pushes that we will try to do, is to get the greatest reliability. And we do that again, as both of us talked about, with redundancy, and the reliability of the components themselves. And we learn a lot, by the way, by past experience. I didn't mention that, but Apollo 14 took off with a lot of things changed to it based on the potential that 13 had. They looked at all sorts of things before they launched Apollo 14, in about a nine-month cycle before they could relaunch it.

QUESTION: I think there is a critical point, that is, evaluation of risk could be approached objectively by a variety of techniques, and you can try your damnedest to reduce that. But at some point someone has to make a decision—.88 or .89, who makes the decision and on what basis do they make the decision? Is it subjective, objective, a democratic vote? How do you do that?

JAMES LOVELL: I'll answer again. The decision is made on, what is the reward? I've mentioned that critical thing on Apollo 8. The Americans thought, our NASA folks thought, that the reward during Apollo 8 was well worth the risk, whereas the Soviets thought that maybe they should send another unmanned spacecraft before risking a new cosmonaut. And, so, you had to look at the reward. If the reward is tremendous, then we have to accept a tremendous risk that is involved with it. Like any other risk factor, if you invest a lot of money in something, you have to think that that's well worth the risk to invest that money to get the reward back. And it's the same way with almost any operation that we do. That's the way I look at it.

MILES O'BRIEN: Jack, just to follow up on that, to what extent did the polar explorers get specific about the risk? Or was it just all a gut feel?

JACK STUSTER: It wasn't a mathematical exercise, that's for sure. It was highly subjective. But it is a personal equation, and some people are willing to accept more risk than others, and it all depends on what the potential benefit is. If the potential benefit is great, then we were justified in taking a greater risk.

MILES O'BRIEN: Would Shackleton and Amundsen have been good astronauts?

JACK STUSTER: I think so. They would have been good mission managers, because they attended to every detail. For them, there was very little risk, because they had already attended to every detail, unlike others who didn't. But, if you attend to every detail, if you had planned for every possible contingency that you can think of that might occur, reasonably, there's a certain confidence in your ability. It's not really taking a risk. The risk is something out of the ordinary, the weather, something that might come up that you can't really count on. And then, you compartmentalize it, and it's okay to deal with it.

MILES O'BRIEN: Any other questions out there? Yes, go ahead.

QUESTION: My question is, how do you evaluate the reward? And, just as an example, think of Cortez and Pizarro, they would have thought that their expeditions to the New World were accomplishing great benefits for Spain, but we see them as genocidal, wiping out great cultures. How can there possibly be an objective measure of reward or benefit?

JACK STUSTER: Well, I think if we encounter other living beings of some sort, or some other entity, that would be a parallel. But, if it's a matter of science, then you measure the actual importance of your discovery, and it becomes, again, a subjective thing. I think astrophysicists and astronomers might be more inclined to take greater risks than geologists to rescue the Hubble [Space Telescope]. I think it is a personal equation. Am I wrong? Well, what I mean is, if the outcome is important to the individual or to a discipline, then those people are likely to take greater risks. But no one wants to make a rash move, however motivated, however wonderful the benefit might be. He wants to make sure that everything has been covered that can be beforehand, to minimize the risk. But the very nature of exploration makes it almost impossible to predict what you will get in the way of benefit.

MILES O'BRIEN: The very nature of exploration makes it impossible to predict what you will retrieve.

QUESTION: Miles, this may be as much for you as the panelists. We have an interesting juxtaposition of risk taking this week. Burt Rutan is getting ready to fly on Wednesday morning. And he would argue that he has carefully balanced the risk/reward ratio, and he is very proud that he's never lost a test pilot. And he clearly thinks the risk is worthwhile. But NASA is often held to a different standard, because it's the government somehow making that decision, rather than an individual. Government has facilitated Rutan's flight, and has clearly said that if he wants to do that himself, he can. But how does the government take that similar risk? Does it get harder and harder to do?

MILES O'BRIEN: Yeah, is the bar set higher because you're a government entity? I think that's true. If Burt Rutan, as a private entity funded by Paul Allen, wants to do this, I think the level of acceptance that people have over the consequences of that, whatever it may be, I think it's greater. And I don't have an easy answer for how a government agency can accept that same level of risk. But, the other side of it is, you have all these smart people in this room, and a lot of resources that you can bring to bear to try to minimize that risk even more. I mean, Rutan has done what he has done so far for around 20 million dollars. And what is that? That's a NASA study, right? A few NASA studies. [Laughter] But, nevertheless, it is pennies on the dollar compared to the amount of cost and amount of resources that NASA has. So, I think that maybe that allows you to accept and create risk with more safety.

JAMES LOVELL: Let me answer that if I could, because I think the classic example was, President Kennedy got up in 1961 and, in his speech, he said we plan to land a man on the Moon and bring him back from space before the end of the decade. Now, we had just put Alan Shepard up two weeks before, in the suborbital flight, had not yet put anybody up in Earth orbit, much less thought about sending anyone to the Moon. So, he saw that this was a risk, because of the position that the country was then in, we were behind the Soviets at that time in technology, and they were doing all these things. And he had to get a position where he thought he could make a bold move. So, he, as the President, represented the government, represented the people, thought that we could do that particular job. It was a huge risk. If we failed, what would be the situation? He took it upon himself as the leader to put us in that position.

JACK STUSTER: Burt Rutan is reducing risk, it's my understanding, to win the X Prize. It is the weight equivalent of three people. There is only the one person and then the weight equivalent of two people that are going up. So, he has reduced the exposure to risk by taking the weight equivalent.

MILES O'BRIEN: I'm trying to get one of those seats.

JACK STUSTER: I know. [Laughter]

MILES O'BRIEN: And there would be no shortage of volunteers, either.

QUESTION: After the Apollo I fire, the Nation grieved. The spacecraft was fixed, and it flew in a very short period of time. After your mission, there was a similar sort of thing. There wasn't a lot of discussion. There was some, but not a lot, of asking, "Why are we doing this?" There seemed to be a compelling thing drawing us out there. Flash forward to the *Challenger*. There was a lot of hand wringing. It took a lot longer than people thought to fly again. Flash forward to *Columbia*. Although something came from this in a space policy, it still seems to be so much more difficult to get back to what seemed to be so natural in the '60s. For either of you, have we lost something since then? Is there something that can be regained? Is there a magic phrase or something you can do that can bring that back, or were we just lucky at that time?

JAMES LOVELL: Well, I think you have to look at the accidents in the context of which they occurred. With the Apollo I fire, there were problems because we did not really understand the use of sixteen pounds per square inch oxygen in ground tests, which we learned very belatedly. The program, at the time in which that fire occurred, was one of intense competition. It was intense prestige in this country. It was one that wanted to be continued to go on to completion. After the Apollo, as we all know, nationally it was sort of like a ship without a rudder for a while. We had various stages of the Space Station. We tried to figure out what to do. What I first recall is we were going to build a space station, a shuttle, and a transportation device all at one time. We found we couldn't do that. I think that a loss today, a *Challenger* or *Columbia*, as compared to a loss during an intense period, is entirely different. We watch these spaceflights take off on television with seven people involved. It is an instant tragedy when we see something like that happen. Actually, I lost more friends testing airplanes until the *Challenger* accident occurred. We took that loss. As you, Jack or Miles, mentioned, sometimes you become accustomed to certain risks. In test flying, we become accustomed to someone buying the farm occasionally, and we didn't think more about that. We try to learn what happened, and then we try to change the system and to improve the system so it won't happen again. Now there are major tragedies. If we lose something, it is a major tragedy because it represents part of our country.

MILES O'BRIEN: Jack, what was the media response after Scott's team perished? Was there a call never to go back to the Poles?

JACK STUSTER: I don't know. I'm not sure what the media response was. There was a great deal of finger pointing and probably a lot of similar response. It didn't stop the progress of exploration

MILES O'BRIEN: People weren't calling for the end of that exploration necessarily?

JACK STUSTER: No. As a matter of fact, it was ennobled. The heroic death of Scott and his polar party wasn't acknowledged. There wasn't the inquiry, let me put it that way, that resulted in the detailed list of changes that must be made for the next one.

MILES O'BRIEN: Question?

QUESTION: In putting together a team for a high risk mission, what relative emphasis should be put on, on the one hand, technical knowledge/training, and, on the other, personal qualities, like resourcefulness and the ability to control an out-of-control situation?

JACK STUSTER: Does it involve isolation and confinement? A small group? Isolated and confined? Technical performance? You will find that people who are technically competent might be called upon to perform their expertise only occasionally, whereas if they are living in isolation and confinement, they are always interacting with each other. So, the skill you should really select for under those circumstances is getting along with each other, and then ensuring that it's the case by demonstrating it, rather than as a test, by simulations or past performance. Of all the principles of the behavioral sciences, ranking them in order of validity, the best predictor of future performance is past performance. You find people who have been successful cooperators in the past, and you have a better chance that they would succeed. If you are going to go to Mars, 13 Norwegians with some seafaring experience would be good. Just don't pack the blubber lamp, right? It would be a bad thing.

If I could return to the question, the question was, what is the relationship of risk and benefit? NASA is compelled to justify the activity. Often it is the spinoffs—Teflon and so forth. There has been one that I have hoped for a long time. Long duration space exploration will result in bone demineralization. It could be the show stopper. The bones become brittle in the same manner that elderly people's bones become brittle, to the extent that it could be dangerous to the explorers, either when they make their planetfall or, certainly, when they return. There are a lot of very smart people financed by NASA who have been looking into this issue to develop a countermeasure. There are people who take the mechanical approach of stressing the bones to trick the osteoblasts and osteoclasts into leaving the bones alone, and so forth. I have been hopeful that a solution would come. I have just learned that if this pans out, it will be the most monumental spinoff that NASA has ever come up with, and, certainly, will justify all previous research and all future expeditions and research, and that is a countermeasure in the form of a pill to bone demineralization. Everyone has an elderly relative, a grandmother, a mother, who fell and broke her hip and either succumbed as a consequence or the quality of her life was changed. All of us look forward to a future where we will live in fear of falling and breaking a hip. This countermeasure successfully developed by NASA will change all of that and would be, as I said, worth all of the effort that went before and will occur in the future.

MILES O'BRIEN: Sounds like a story to me. [Laughter]

JAMES LOVELL: Jack, you're not suggesting we send John Glenn up again, are you? [Laughter]

MILES O'BRIEN: Okay. That's all the time we have. Great panel.

SESSION ONE

earth

Hunting Microbial Communities
in Dry Antarctic Valley Lakes

I'm going to talk about my experiences in both the Antarctic and in the Arctic, and I'll focus more on the Antarctic; right now it's a little bit more exciting.

We started back in the late '70s, working at some of the lakes that are in the dry valleys. And this is one of the largest ice-free regions that exist there. There is a series of lakes with very, very thick ice cover, and nobody had ever probed these lake before, and looked at them in depth, other than just taking some drills and popping small holes in them, and taking a soda straw approach. We wanted to look at them in a little different fashion.

As I started thinking about this symposium, I started categorizing the kinds of risks that we were involved with over this period of time. And for us it really came down to mission risk, the success of our expedition in general, and personal risk, which I divide into transportation to and from the research site.

We usually take large ski aircraft to get down to McMurdo Sound, and then to get up north we use Twin Otters. And then to get into our remote field sites, we'll take helicopters, or, in some cases, on some of the expeditions we've taken Soviet icebreakers around the

Dale Andersen
Astrobiologist, Antarctic/Arctic Researcher, SETI Institute

Dale Andersen has made almost 1,000 dives beneath the thick ice of lakes and oceans. He helped develop a number of techniques needed to explore the lakes of the McMurdo Dry Valleys, including both human and robotic exploration (using the first remotely operated vehicle—ROV—in the Antarctic). In 1991–92, as a member of the Exobiology Implementation Team under the auspices of the U.S./Soviet Joint Working Group for Space Biology and Medicine, he led the U.S. field team on a six-month joint U.S./Soviet expedition to the Bunger Hills, Antarctica. Currently, Andersen (along with Wayne Pollard at McGill University and others) has been studying perennial springs and ice-covered lakes in the Canadian High Arctic.

continent, and then taken helicopters inland and have been dropped off. We don't have much control over what happens during those times. We're there for the ride and, hopefully, we make it. And obviously sometimes the weather and the conditions can get pretty bad along those lines as well.

Then, we actually live and work in these remote environments. We have to learn how to do both of those things—we have to learn how to live effectively. That takes quite a bit of effort, if you're going to spend lengthy periods of time in these remote settings.

And, then, it's not just living, but you also have to accomplish some work. Developing the skills that require you to do that work is also something that

NOW WHY DID WE GO TO THESE PLACES? WE DID HAVE SOME OF THE PROGRAM MANAGERS WHO WERE ASSOCIATED WITH THE PROJECT TELL US . . . THAT WE WERE JUST GOING TO FIND A BUNCH OF ROCKS, AND THAT WE'D WASTED EVERYBODY'S TIME AND MONEY. WHAT WE FOUND WAS A REALLY LUSH, LUXURIANT, MICROBIAL ECOSYSTEM THAT HAD NEVER BEEN SEEN BEFORE.

we've had to overcome over the years. And then there is also the scientific credibility from a personal standpoint. We can put all the effort into going to these regions to do the studies, and if we get there, and the weather is too bad, or if the equipment doesn't work, or we've forgotten something really important, it's just too bad. And, then, we come back with nothing but our hands in our pockets, and say, well, we had some good times. And, then, the program managers yell at us and don't give us any more money. The ultimate currency would be scientific publications. And if you don't get those scientific publications, then you don't get to go back down and do that kind of work again.

And, then, there is the family setting. It's always difficult to be away for these long periods of time, especially when you are in a hostile environment. Your family doesn't know from a day-to-day perspective if you are just wandering around in the snow or if you're drowning underwater or whatever is going on. So that's one of those long arms that's always reaching out to you.

Over time, we've really had poor communication. And it's something that has only changed in the last five or six years. Now we can pretty much have global communication on an instantaneous basis. It used to be that we just didn't have any; it took weeks for something to get through.

Then there's risk to others, or assigned risk. If I'm a leader on an expedition, and I can send people out to do work, my level of risk assessment is different than what I would expect for somebody else, depending on their level of training or what their experience is.

For a program manager, that perspective might be completely different. It's different if they've never been in a field situation similar to that, or if they have any kind of mountaineering or diving experience, or some other kind of experience that they understand personal risk from that standpoint. But it's valuable to have experienced people in those positions.

Again, there are different kinds of transportation risks that I've been involved with: large aircraft, large icebreakers, even Hägglund track vehicles that we can take across the ice shelves among the continental ice, and helicopters. All of those have different types of risk associated with them. And then in some cases, like the Hägglund track vehicles, we drive those personally, so then we have to be able to understand how to drive a vehicle like that, keep it out of crevasses, and understand what to do when it breaks down, which it invariably will.

In the dry valleys, a lot of the ice is not in the valley itself. You have the continental ice flowing through, and you have a barrier that essentially keeps a lot of this from going into the valley. The evaporation highly exceeds the precipitation, so it's quite dry. But some of the small features in between the glaciers are actually lakes. One example is Lake Hoare, where we spent quite a bit of time studying the lake ecosystem itself. Now these lakes are permanently covered in ice, and the ice thickness ranges from about 3 to 4 meters up to 6 meters. In earlier years, about the only way people studied these lakes was to go out and drill a small hole through it, and then take water samples and take some other measurements. But nobody ever really asked the question, what's going on at the bottom? Because most people thought the light levels were so low that there wouldn't be anything on the bottom.

Well, a few years later, we decided to open these lakes up with a bigger hole so that we could go down and look. I guess, for some reason, we weren't smart enough, or at that time underwater cameras were still a hassle to get a hold of, and it was just as easy to put us in the water as it was to take a little camera and drop it down the hole. But that wouldn't have had nearly the fun factor, either, I guess.

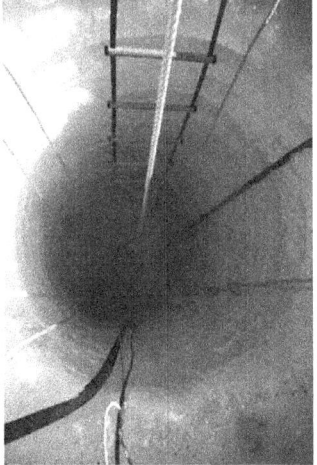

We developed a system that allowed us to melt holes through this thick ice. Essentially, we take a copper coil and just put hot antifreeze solution through it, let it melt down the ice. You can see that we can move in and out of that ice cover, as we're melting it. It's like a tunnel, of course, a water tunnel. Then we suit up and we jump in. Originally, we started out in wet suits and double hose regulators, which is kind of ancient history. Over the years, that was obviously not the best solution, because you've got to deal with a really cold wet suit once you get out of it. If it's minus 30, it's not too comfortable to take a wet suit off in those conditions. You can see that we're on tether, because once you go in, you've got one way in and one way out.

Now why did we go to these places? That to me was the biggest payoff. We did have some of the program managers who were associated with the project tell us, just before we were getting ready to go in for the first time, that we were just going to find a bunch of rocks, and that we'd wasted everybody's time and money. What we found was a really lush, luxuriant, microbial ecosystem that had never been seen before. And as it turns out, these tunnels that we make through the ice are just like tunnels back through time, 600 million years to 3.5 billion years ago. This gives us a glimpse of early Earth, and, perhaps, of early Mars. If lakes existed on Mars, they would possibly have been ice-covered, and possibly have microbial communities similar to what we see in the dry valley. These kinds of microbial mass communities are very, very prolific.

Our work there is very dependent upon the training that we received. We have to be very safe. We're diving in extremely remote areas where we don't have the opportunity—at least, at that time, in this particular area, we didn't have the opportunity—to get back to a recompression chamber. The closest one would have been in New Zealand, and that's, of course, after a helicopter possibly came out, and picked you up, and dragged you back to McMurdo, and put you on a C-141 Starlifter back to New Zealand.

When we were in other situations, we didn't have that option at all. The nearest chamber was probably several months away. And we worked everywhere from about 40 meters up to the water columns. We have to be very, very careful about the dives that we go on, and the kind of diving that we're doing. That one tether that takes you back to the surface is so very important. So it's kind of like "follow the light." But it's actually very fun diving. It's very exciting to go into these lakes.

We also started a program a number of years ago. Because we can only go so far out on our leash underwater, we really are restricted to the amount of material we can see. So we started a telepresence project where we started using remotely operated vehicles—actually Sylvia Earle loaned us the first ROV [Remotely Operated Vehicle] that was used down in the Antarctic in one of these lakes. A lot of the algorithms that we used for one of our ROVs was actually utilized later on the Mars Pathfinder. Some of guys in the Intelligent Mechanisms Group at Ames worked very closely with us on that. Anyway, that was a very good means of getting out and away from that dive hole, and getting into working with other communities. We also had to learn to operate this kind of equipment in those environments.

Now we just don't follow the microbial communities underwater. We also follow them wherever we can find them. That includes on top of the glaciers. We then have to pick up ice climbing skills and some general mountaineering skills when we want to go up into the alpine areas. So, it's a real skill mix to work in these kinds of regions.

In 1991–1992, I had the opportunity, along with Jim Rice who is here in the audience and Peter Dorn who is in Chicago, Illinois, to go to the Bunger Hills, another ice-free area in the Antarctic, with the Soviets. That was a joint expedition between our two space programs through the Space Medicine and Biology Working Group, and it was part of the Exobiology Implementation Team that allowed us to get together with their exobiologists.

Three of us joined eight Russians. We traveled for two months around the Antarctic continent by icebreaker. We got dropped off by helicopter for the next four-and-a-half months in extremely remote settings with quite a few cultural and language barriers. To be honest, the expedition took place at the very last minute because of the way programmatics worked and the way permissions go, and because the dialogue between the Soviets and the U.S. at the time was kind of slow. We didn't find out that we were actually going until four weeks before we left, so we had to buy everything, ship everything, and get everything down to South America. Hopefully, it was all there when we got there. We actually didn't open up our boxes until we, literally, got to the Bunger Hills and the last helicopter left. It was a pretty quaint setting. We spent some time in some of the huts that they had there, but we also spent quite a few months sleeping in tents on the ground in the rocks. There was a great deal of wind. This is an area that is prone to katabatics.

So we had quite a few difficulties while we were there, but we also had to accomplish some work. All in all, it was a very successful mission, both scientifically and culturally. It was culturally enriching to all of us, and it showed, at that time, even though it wasn't a space project per se, the two space programs were working closely together in an isolated, Mars-analogue environment.

More recently I have shifted poles, and we are now working in the Canadian high arctic. The setting is not altogether that different. It is very glaciated. There are large Alpine glaciers and outlet glaciers, along with large ice sheets. We're studying a series of perennial springs in this region. These are the highest latitude perennial springs in the world, along with those over in Svalbard. This is a great Mars analogue. There may have been springs on Mars coming up through thick, continuous permafrost. This is where we would go to see these. Again, this is a very remote setting to which we take Twin Otters and helicopters, are dropped off and left for a few weeks or months. It's a little shorter logistics train up to the Arctic, so we can get back and forth much easier. Nevertheless, while we're there, we are actually a little more isolated, at times, than we are in the McMurdo dry valley area.

High-Altitude Mountaineering

I've been climbing most of my career and, hopefully, I've done it the right way. I started by climbing small peaks and then aspired to climb higher and higher peaks. As I was going higher, I eventually realized that what I liked about mountaineering was the extreme challenge and how difficult it was and, also, the beauty of climbing these mountains without oxygen—the pure way, the hard way. I had probably a difficult start, though, in my career, being raised in the great mountaineering state of Illinois. But it was in high school that I was reading adventure books: Amundsen, Scott, *Endurance*, and books like that. I came across a story written in 1950—*Annapurna*, and it details the first ascent of an 8,000-meter peak. There are fourteen 8,000-meter peaks in the world, roughly 26,000 feet and above. In 1950, a French team climbed this peak called Annapurna. They succeeded, and it is a very amazing story. It inspired me to start climbing mountains.

The first thing I had to do, obviously, was to get out of Illinois, and I chose to move to Seattle. My other goal in life was to become a veterinarian, so I studied veterinary medicine in the Seattle area. I started climbing voraciously and learned from people who were very experienced, very conservative, and willing to teach me their craft. Eventually, I landed a

Ed Viesturs
American High-Altitude Mountaineer

Ed Viesturs is America's leading high altitude mountaineer, having climbed many of the world's most challenging summits, including ascending Mount Everest five times. He holds this record with one other person. Viesturs is the only American and one of five people to climb the six highest peaks in the world—all without supplemental oxygen. He is currently on a quest to climb all 14 of the world's highest mountains (above 8,000 meters) without the use of supplemental oxygen. He has successfully climbed 12 to date.

job guiding on Mount Rainier, 14,400 feet, which I have now climbed 192 times. Most of those times were as a guide.

I truly believe that my safety and success in the Himalayas is because of my guiding experience, because, as a guide you always have rather inexperienced clients in tow. You are responsible for them. You always have to be evaluating risk and always asking yourself, if this happens, how do I get out of that? So you are always trying to think ahead, or, at least, not waiting until something were to happen. I truly think this has helped me a lot in the Himalayas.

As I was guiding and climbing higher and higher peaks, I started dreaming big. Obviously, the highest peak you can dream of is Mount Everest at 29,035 feet. I thought that if one day I had the chance to go there, it would be an amazing thing. On the summit there is one-third the amount of oxygen that there is here at sea level. To date, almost 2,000 people have stood on the summit of Mount Everest. The kind of clothing that we wear is very lightweight. It's very technical. The boots are thermoplastic. There is foam insulation. The fabrics we use are single-layer GORE-TEX®, breathable, waterproof, and these products insulate us, but we still have to move to create body heat. We don't have any internal heat source other than our bodies. Basically, these products do not help us climb the mountain, but they make it safer. We climb faster and more efficiently.

Most people that climb Everest—I'd say 95 to 98 percent of the people that climb Everest—use supplemental oxygen. Even with supplemental oxygen, it is very difficult to climb this mountain, because the oxygen is mixing with ambient air, and you are only really reducing your relative altitude a couple thousand feet. It is still very difficult, but I decided long ago that if I ever had the chance to go to Everest, or one of the other 8,000-meter peaks, I would not use oxygen to climb these mountains. I thought it would be more interesting to challenge myself and to see if *maybe* I could get to the summit, rather than guaranteeing myself getting to the summit. So, that is a rule that I made long ago and something I've lived by.

I had my first chance to go to Mount Everest in 1987. We were climbing the North Face. With my partner, Eric Simonson, we were part of a large team of about 10 climbers, and it takes 2 months of climbing, setting up a series of 4 camps, and carrying loads to just get into position to go to the summit. In the end, if you can get one or two people from that team to the summit, then that is teamwork, and that's a success. So, Eric and I were making the final dash to hopefully get to the summit. I was climbing without oxygen the last 300 vertical feet of the summit ridge to the top of the world. It was kind of late in the day, and the weather was changing. Eric, rather conservative, as am I—we were looking at the top, thinking that we could probably get ourselves to the summit, but we probably wouldn't get ourselves down. For both of us, that seemed to be a huge decision-making factor. I have always felt that climbing has to be a round trip, right?

But a lot of people lose sight of that. They see this goal that's two hours away and they've spent years of preparing and training and months of climbing, and when it gets to that close of a distance, they're willing to throw caution to the wind simply to get to the summit and then, hopefully, they'll get down. And a lot of those people, unfortunately, never make it back down. For me, the risk is too great—I don't want to die going to the summit of the world. So we walked away—300 feet from the top. And it's something I thought about every single day for three more years.

Two years later, in 1989, I climbed the third highest peak in the world, Kangchenjunga. We climbed by a very steep and difficult route—this is the technical climbing that we had at 24,000 feet. Very strong team, rather good conditions, and in the end, we reached the summit of Kangchenjunga at 28,200 feet on a very, very pleasant day. And if the weather's good, and you have time, you can stay for about an hour on the summit. If the weather's bad, three minutes, take some photos, and down you go. We stayed for an hour. Looking 80 miles to the west, you could see three more peaks in the distance. And the one closest to the center, the large one, is Everest. And even though I was there standing on Kangchenjunga, I was longing to be back on Everest to finish that last 300 feet.

I HAVE ALWAYS FELT THAT CLIMBING HAS TO BE A ROUND TRIP, RIGHT? BUT A LOT OF PEOPLE LOSE SIGHT OF THAT. THEY SEE THIS GOAL THAT'S TWO HOURS AWAY AND THEY'VE SPENT YEARS OF PREPARING AND TRAINING AND MONTHS OF CLIMBING, AND . . . THEY'RE WILLING TO THROW CAUTION TO THE WIND.

A year later, I did go back to Everest. And—we were making the final ascent, 300 feet away from the summit once again. Climbing without oxygen at these altitudes, you go very, very slowly. You take a step and you breathe 15 times. You take another step and you breathe 15 times. And then you think about taking another step, and you breathe 15 times.

But after 12 hours of continuous climbing from the high camp at 27,000 feet, I then, finally, stood on the summit of the world. And to me, this was the most amazing point of my career—to be on the top of the world. I've always told people this is the closest you can get to outer space without actually leaving the ground. This was an amazing moment for me.

The following year, I went to K2—the second highest peak in the world—and by all means, much more difficult to climb than Everest. It's steeper, the weather is worse, and only 200 people to date have climbed K2. And I went there with my great friend from Seattle, Scott Fisher.

At 26,000 feet was our last camp before we made the final dash to the summit. The shelters and tents weigh about five pounds. On K2, the higher you go, the steeper it gets and the more dangerous it gets. And that's where you really,

really have to be careful, and that's where a lot of accidents occur, because people are so focused on just getting themselves to the top, they're not even aware of the weather, their surroundings, and the conditions. And that's something that you really have to evaluate every step of the way. And, luckily, we climbed it on the first attempt. And once you climb K2, you never think of going back.

I also met that year on K2 another great friend, Rob Hall from New Zealand. We became great friends and did many climbs throughout the world. We climbed Everest together several times as well as Lhotse, Everest's neighbor. Lhotse is the fourth highest peak in the world. And what we started to do was to do tandem climbing where we would go and climb Everest over a two-month period, acclimatize our bodies, and then, quickly, in succession, climb another 8,000-meter peak in just three days. And so we could utilize our acclimatization.

Tragically, though, both Scott and Rob died in 1996 on Mt. Everest during the tragic storm of May 10th. I was there with my wife. I was the climbing leader for the Everest IMAX film. My wife, Paula, was our base camp leader. Our expedition leader, and the director of the film, was David Breashears. Very experienced climber, very detail-oriented, and an amazing leader to have on this climb, where you had a very difficult task in taking a 42-pound camera to the summit, doing what other climbers were doing but also schlepping this giant microwave-sized piece of metal to the summit as well. And after six weeks of climbing, we were in position to go to the summit. We were going to go on May 9th—this was a day ahead of everyone. We evaluated the weather conditions. The weather patterns weren't what we were hoping they would be and we decided to go down. Our friends, led by Rob and Scott, made the decision to continue on to the summit on May 10th and then, sadly, the storm occurred, and on that day eight people died—two of them my great friends. There was nothing we could do.

We couldn't climb up to them fast enough; we couldn't get a helicopter to get up to them. And that's the thing about mountaineering. It's one of the places on the planet where rescue is literally impossible, unless you send humans to go do the rescuing. There are no machines and there's no other way to get these people.

Part of the reason, I think, that the tragedy occurred is that people were swayed by the group decisions that were made. It was almost like, if six people are going, then ten people are going, then twelve people are going. You know, there's comfort—there's psychological comfort—in groups with larger numbers. And I think people would say, "Well, they're going, so I'm going to go." And then they kind of just kept pulling each other up higher and higher late in the day as the storm was brewing. We helped with the rescue. We quit filming and stopped what we were doing in relation to our own objectives. We helped bring some people down and then, after that, we managed to pick up the pieces, and we went back up the mountain two weeks later. Not in spite of what happened, but, I think, to show the world that you can climb these mountains and live to talk about it. If you find the right conditions, and if you wait with patience for the proper weather, you can climb these mountains and live to talk about it. The summit is only half the trip, and the most important half of the climb is, in my opinion, the descent.

I've gone on to climb more. I'm on a quest to climb the 14 8,000-meter peaks in the world without supplemental oxygen. And over the last 15 years, I've managed to climb 13 of them. I have one more to go, which is called Annapurna, the mountain that got me into all of this. So maybe it's poetic that it happens to be the last mountain on the list.

Exploring the Deep Underground

I've been trying to get to Mars ever since I was a little kid. And my solution to this over the last decade or so is, really, to go down instead of go out. And that may be as close to going to another planet as I ever get. After we published a paper with Chris McKay suggesting that life on Mars would be underground, we were looking for ways to study the sub-surface. And, of course, the most immediate and obvious thing was the new information we were getting from drill holes. People were beginning to drill, through the Department of Energy project, and trying to look at deep subsurface microbiology. We discovered that caves were there and that maybe they would be cheaper to get into. And so, after the early part of my career working in extreme environments on the surface, I decided to try caving. The first cave I ever caved was Lechuguilla Cave in New Mexico, which is a notoriously difficult cave. And all I thought at the time was, "Am I going to live to get out of this? I just have to live to get out of this cave."

But after the pain sort of faded, I realized that it was an amazing environment, that I'd never done anything in any place that was so potentially fascinating for the kinds of exotic microbiology that I was interested in as an astrobiologist. And so I realized that I could do

Penelope J. Boston
Director of Cave and Karst Studies, New Mexico Institute of Mining and Technology

Penelope Boston's areas of research include cave geomicrobiology, microbial life in highly mineralized environments, and unique or characteristic biominerals and biosignature detection. In addition, she is involved in astrobiology and the search for life beyond Earth. At New Mexico Tech she is creating a new program, Cave and Karst Studies, in conjunction with the founding of the National Cave and Karst Research Institute. She is extensively involved in educational outreach about caves and karst issues, space exploration, and general science for school children, older students, and the general public.

two things. One is, I could learn how to cave safely and I could go to these places. And I could then refocus my work to essentially tap into an entire area of biology and mineralogy, and the way they overlap on our planet, that had not been studied before. This is a new field, really, in terms of what we've been doing on Earth. And then it was immediately applicable, of course, to the situation of life beyond the planet.

And, so, most of my research now is focused in one type of cave or another. We tend to pick them for their specific chemical properties, so we're looking for caves that have poisonous atmospheres, that are very hot, that are very cold, that are very extreme in some sense, so that we can look at the limits to life on this planet and learn from that what may be adaptive strategies of life for other bodies in the solar system and, perhaps, someday beyond. So really, to write a field guide to unknown organisms is part of our mandate.

I'm studying the caves that we are looking at as I'm building on the work of other cave scientists in the world. There are a number of really clear lessons that we're learning. These amazing environments show us that caves are really not rare. If you have only ever been in caves as an occasional tourist in a show cave, you may think that caves are a rare phenomena, but really, there are a tremendous number of sub-surface voids on Earth of all different kinds. They aren't just in calcium carbonate-type environments, which are the ones that we often come in contact with, but they really occur in every major rock type. And this is an important lesson for trying to apply our knowledge of Earth caves to other bodies in the solar system.

There are many, many ways to make caves. And, so, one of the areas of active research that we're engaged in is, really, a set of thought experiments about how you can look at the basic physics and chemistry of environments and try to imagine ways that subsurface voids on other bodies could be formed. And, so, we're working that end of the theoretical spectrum of imagining caves on other planets.

The type of cave that we absolutely know exists elsewhere in the solar system are what are known as lava tubes, and these are a natural outgrowth of flood-basalt type, quiet, flowing lava eruptions. And these things are essentially rivulets that freeze on the outside. The rock on the outside freezes. It's a very good insulator, and then it allows the interior to remain molten and lava continues to flow through. Eventually, when the eruption stops, these empty out and you have these beautiful tubes, and so that's a very different class of cave from the kinds of dissolution-dominated caves that we often think of.

Certainly, it was known and recognized by Ron Greeley and other colleagues, even in the Apollo era, that a lot of structures that they were seeing on the moon were lava tubes, or unroofed sinuous rills, which is like a lava tube without its top. As we get ever better imaging of the planet Mars, we have seen that there are lava tubes scattered widely over the planet, and these are quite easy to pick out. Olympus Mons has this little pit crater or little pit collapse feature, a string-of-beads kind of appearance. This is a direct analogue to the way we find these sorts of things on Earth.

One of the interesting things that's been attributed to the fact that the gravity on Mars is much lower is the fact that the lava tubes scale accordingly. So not only does Mars have enormous examples of volcanism, but it has big lava tubes. The biggest lava tube on Earth is about 90 kilometers long, in Hawaii. That's the record-holder on Earth. But, typically, when you look at these features on Mars, they're hundreds of kilometers long. And the diameters are equally great, so they're on average 3 to 10 times the size of the average diameter of a tube on Earth. So these things are truly enormous.

Not only those places, but when we look at the radar imaging data from the Venus missions, you can see that there are tube-like structures associated even with those weird-looking types of volcanic features that you find on Venus. And,

IN A LOT OF THESE SUBTERRANEAN WORLDS, WE REALLY ARE THE

ALIENS, AND THIS IS AN IMAGINATION-STRETCHING EXPERIENCE.

then, even Io, which is such a cooking little moon out there with its tremendous sulfur component, seems to have what is clear evidence of lava tubes. And my dream is that somebody will get a really good image of one that's made out of entirely molten sulfur.

So these are fabulous features. These are places—at least on the Moon and Mars, although I wouldn't recommend astronauts going directly to Io or Venus— that can actually be exploited as human habitat. We just finished a Phase 1 and Phase 2 study over the last four years for NIAC, which is the NASA Institute for Advanced Concepts, looking at the far-future conceptual development, and looking at ways to discover the likely enabling technologies that we would need to make these actually useable for structures for astronauts and bases on the Moon and Mars in the future.

So, I'm not going to talk about all the different kinds of caves because that would be a several-hour lecture in and of itself. But one of the points that I think I want to really press home is that cave environments are typically radically different from the surface. Pictures taken in Saudi Arabia by the very well-known caving team of John and Suzy Pinch have shown that, even in these very hot, blasted sand deserts, when you get into very large bell-shaped caves, there are divable pools. And the air in these caves is near saturated humidity. So it's a complete change from the overlying environment, even in caves that are not sealed. And, so, it's just the barrier of above and below that provides this radically different environment. And this is a big message for astrobiology, that just what might be dominating on the surface of a planet is not necessarily the key to where you actually have to go to look for the life.

Cave environments obviously have no sunlight, so this means that any organisms living within them have to make their living elsewhere, either by detrital organic material washing in or, in the case of a lot of the organisms we're studying, they're essentially rock eaters. These guys are disaggregating the parent rock using things like the metabolic product, the organic acids that they give off, and, then, other organisms come along within these little microcommunities, and they oxidize the metals in the rock. And this is how they get the energy to run their entire ecosystem.

They're very high-humidity environments. In contrast to the surface, they're very thermally stable, so even a cave with a big, gaping, open entrance still remains very, very thermally consistent on the interior. New transfer is usually very, very low, with some exceptions. They're very rich mineral environments. And then there's no conventional weather, so it very much is a very different planet in the near-crustal caves than it is on the surface.

And the result of all these tremendously different conditions is that caves are unique mineral factories. There are vast numbers of unique mineral formations that are found in caves. Huge catalogues of them have been seen and the explanations for the occurrence of these is very much in its infancy. One of the things that we are working on extensively is which of these types of mineral formation processes are actually biogenic, and it turns out that there are a lot of them—perhaps even most of them in caves have a biogenic component.

So the organisms that are actually contributing to what's going on in caves are not simply passive observers or users of the environment. They are mineralogically interactive. They are changing the caves. They are actually interacting with the bedrock and they are guiding, and in some cases controlling, the kinds of mineral deposits that are left. A lot of these organisms are novel, and I would venture to say that the bulk of the organisms that we find are actually novel. They're not known to science. And, so, from one little cave puddle to the next, perhaps we have 80 percent novel organisms that we look at by molecular phylogenetic techniques. So these are truly evolutionarily, self-contained environments, and many of them are actually physically isolated from the surface, and, in that sense, they really are miniature planetary systems within our own crustal environment.

Not only do the caves house this amazing array of organisms that we're just beginning to understand and study, but also they're wonderful preservation environments. So if you are looking for biosignatures, then the subsurface in caves is the place to look. Not only do the organisms live there, but they very often self-lithify. They're engaged in self-fossilization while they're actually alive. So-called U-loops from Lechugia Cave look very organic. They are entirely rock now, but we have been studying their living counterparts in modern caves, and these are clearly the fossilized remains of microbial mats that were inhabiting this cave probably on the order of four to six million years ago when this cave was actively forming. We can also expose the fossils in this kind of material by acid etching.

In a lot of these subterranean worlds, we really are the aliens, and this is an imagination-stretching experience. It's the kind of thing where it would be lovely if we could take a lot of program managers into this kind of environment, because, really, it's the kind of thing where just reading about it doesn't make as much of an impact as if you were actually doing it.

Cueva de Villa Luz is one of the most amazing caves that we're studying. It's a sulfuric acid-saturated cave in Tabasco, Mexico. Gases from the nearby volcano, El Chico, now actually come into this cave and make it an extremely poisonous environment within which to work—tremendous amounts of hydrogen sulfide, carbon monoxide, carbon dioxide, even aldehydes, and various other noxious things. And so we require complete protection from that environment.

The message here is that this cave is the most biologically-rich cave of any that we've ever seen, and it's because of these poisonous gases. These poisonous gases are not poisonous to the organisms that are living there; it's home sweet home. And this is the message. We're not looking at extreme environments just to look at extremes where organisms are just barely hanging on. We're looking at them to look for organisms for which that is the comfortable environment, because those are representative of what we may find as the average conditions on other bodies.

So we're trying to write the field guide to unknown life, and this is a really tough thing to do. But the place that I've been in my career where this makes the most sense to me is in these kinds of protected and evolutionarily-sequestered environments. A lot of the material we look at doesn't even look alive. We've seen white muddy-looking stuff on the walls—that's living mud. That material is made out of cells and filaments that coat themselves with calcite mineral. And they're actively producing this material in caves all over the world. Tiny white dots on the walls are organisms that are busy dissolving the salt in a lava tube and making their living there. So, even though something may not look alive, and although sometimes we have to work very hard to show that it's alive, all of these environments contain amazing life forms that are busy also leaving traces of themselves.

One of the other aspects of doing the kind of cave work that we do is also giving us operational experience that is very valuable to future life detection missions, whether they be robotic or, ultimately, crewed teams in the future. And that is that we are operating in extreme environments that are hazardous, with an indigenous, sensitive, alien biology. In this case, the alien biology is on our own planet, but, nevertheless, it's very different from our surface environment, and we have to take all the precautions that one would imagine in order to avoid contaminating it while, at the same time, managing not to kill ourselves off while working in very difficult conditions. And so, it is an analogue for operating with life-detection constraints, including even the aspect of working with various collaborators at MIT [Massachusetts Institute of Technology] and JPL [Jet Propulsion Laboratory] and NASA-Langley on robotics that can get us into some of these kinds of environments.

So the caves are out there. I know that as time goes on and we explore the rest of the planets in the solar system, we'll find better and better ways of actually detecting caves besides the lava tubes. We'll find ways to get into them. We'll find ways to drill into them—which will be a lot easier than just sinking a core right down into solid rock—and they will have amazing structures, amazing minerals, and, perhaps, even amazing life as we explore them.

NASA Image/Renee Bouchard

Exploring the World's Highest Lakes

I am a planetary geologist, so I am revolving between Earth and heaven, and part of heaven is Mars. As such, I am very privileged and, also very proud, to be part of Mars Exploration Rover science team, which is proving a lot in terms of exploration. And, as a member of the science team, with my colleagues, we are taking intellectual risks. We have been working for several years to select landing sites, and we tried to match the landing sites to the objectives of the Mars Exploration Rover mission. We were very happy to see that, indeed, in many cases we have been able to reach our objectives.

But, although it's an intellectual risk, it's also a risk for the assets and for the time that has been put into preparing these missions. There is no human life involved in that part of it, but still, you want to make sure that you are doing your job properly. And to do that, you have to do your homework. I would volunteer right away to go to Mars. People here, like Mr. Lovell and [Harrison] Jack Schmitt and others, made me dream as a little girl, and I knew that whatever I was doing with my life, it would involve NASA, planets, and exploration. But still, you have to make sure that the concept you are pushing forward is as close as possible to reality. And to do that, we had to do a lot of planetary geology here on Earth and to try to find the best analogue we can.

Nathalie Cabrol
Planetary Geologist, NASA Ames Research Center, SETI Institute

Nathalie Cabrol is a planetary geologist at NASA Ames Research Center, since 1994, and a SETI Institute principal investigator, since 1998. Her research focuses on aqueous environments favorable to life on Mars, exploration (robotic and human), and the study of terrestrial analogues. Her education includes a master and doctorate degree in planetary geology (1986 and 1991 respectively), both obtained at the Sorbonne University in Paris.

Nathalie enjoys climbing and hiking. She also practices free diving (diving without oxygen tanks). She holds an (unofficial) women's world record for the highest free dive, performed in 2003 in the Licancabur summit lake. Both her passion for mountaineering and diving are combined in her scientific research for the limits of life on Earth.

I am interested in lakes. I am interested in past aqueous environments on Mars and their habitability potential—whether those are the best place for life, whether life ever appeared on Mars—these are my drivers as a scientist. And to try to understand that, I am exploring high lakes. Why explore high lakes? Because the higher you go on Earth, the earlier you go back in time. On Mars, basically, you will be going into a core environment temperature that is really low and you are going into a thinner atmosphere and you try to understand what is happening over there. And, so, we are exploring those high lakes. We have been starting in the Andes with volcanic lakes.

And our goal is astrobiology, but as we were going, our exploration kind of caught up with us, and we discovered a brand new avenue of research. And this involved physiology and medicine—and I'll talk a little bit more about that—and, of course, education and public outreach.

Well, why go high? As I said, these are the best analogues to Martian lakes and we want to understand the limits of life. And why do we want to do that? Because right now, the Mars exploration rover mission has proven beyond doubt that Mars was a habitable planet for the type of life we know here on Earth. But that does not mean that there was life there; habitability and actual life being two different things.

So, going back to these high lakes is for us to understand: Is life possible in analogous conditions to those on Mars? So we are going up there, and we are trying to start with crater lakes in the Andes, and we'll try to move on to the Himalayas, because as the lakes in the Andes are slowly receding now because of the climate change, by the same token, on the other side of the planet in the Himalayas, large glaciers are now melting and they are creating new lakes that are probably higher than those of the Andes. This is, for us, a place to really stop and witness the ecosystems forming: How life gets there and develops, et cetera. So this is really something that's fascinating.

And we started with the Andes and, well, you have to imagine, about two years ago, myself entering the office of my branch chief at Ames and saying, "Hi, Chief. I think I have this very bright idea. I would like to climb a 20,000-foot volcano, named Licancabur, at the boundary of Chile and Bolivia. By the way, it's not extinct. And there is a lake on top, and I want to dive on that lake. Of course, it's almost freezing temperature, but I forgot to tell you something, I am a free diver. I am not using oxygen. Can you help me with Code Q [NASA Safety Office]. He actually did, and this is how we started the Licancabur adventure.

We were talking this morning about risk and payoff. It has been the most rewarding experience of my entire life so far. We were blessed by the fact that nobody in our crew was hurt, nobody was really in significant trouble. You will wake up lousy at that altitude, because this is the kind of thing that happens, but nobody got actually sick from mountain sickness. But we had to work with Code Q for six months. And among the people who advised us on this trip was Peter Hackett, and Peter was one of the other guys climbing Everest without oxygen. So we took this very seriously.

But, in the meantime, as is part of exploration, you open one door and many, many others open as you are walking. And so we started climbing Licancabur, trying to determine if there was life up there, and as the title of our project was "Exploring the Limits of Life in the Highest Lakes on Earth," I had never imagined that that could be testing our own limits. And not only were we writing the experiment, but we became part of the experiment. And this became fascinating.

We know how analogous these places are to ancient Mars. The pressure is the same. The temperature is the same. UV radiation—well, according to models, we shouldn't be very far off on this one. It's a very arid environment, yet the caldera is right next to it. We're in a volcanic environment with variable thermal input, and if we have questions about life on Mars, when we go to these lakes, we discover that life is thriving over there.

And so, to us, that was very, very encouraging. Because this opens doors and potential for the study and the search for life on Mars. And going up there normally would help us to understand better from the standpoint of the biology, but it helps us, also, to prepare for a future mission on Mars, because we have a better idea of the type of instruments, of the type of exploration strategy, what we would be able to prepare, and giving heads up to management. And this is very important, too.

AND WITNESSING MY FATHER'S LAST DAYS IN THE CONDITION HE WAS IN, AND THE DIFFICULTY HE HAD BREATHING, I SAID THAT THE ONLY RISK I COULD TAKE WOULD BE TO HAVE ONE MORE PERSON TO SUFFER THAT THING.

Our summit camp was around 19,500 feet, and we could look down to two evaporation lakes, Laguna Verde and Laguna Blanca. These two lagunas are lakes which were really perennial lakes 15,000 years ago during the last glacial maximum. And since then, because of climate change, they have been evaporating. And we used those two lakes, which are at 14,000 feet, as an acclimatization area. And we are studying them.

Once again, we stumbled into the unexpected, which is the sense of exploration. And as we were just hoping to spend a little time there, to prepare for the ascent, then we started discovering that we had more analogy with Mars than we had bargained for. Everyday at about eleven o'clock, because this is a thin atmosphere and because you have big gradients of temperature due to the high volcano, you will have huge dust devils roaming around those lakes, and they will engulf your tents and your refuge. And you better not be in the middle of it. So that also tells us about human exploration, that there are things you must be aware of.

But what we also stumbled into were stromatolites [fossilized algae colonies, some of the oldest and continually surviving life forms on Earth]. And the field we stumbled into is 150 square kilometers. And these are on the paleo shore of this larger lake I talked about, Laguna Verde, Laguna Blanca—15,000 years ago this was one lake. And these are fed by hydrothermal springs. And we discovered those stromatolites. When we examined those structures, we discovered that they were biogenic. This means that we had blue and green algae forming then, and we know that these microorganisms are the ones that existed on planet Earth. And you have to go back to archaea [single-celled organisms, among the earliest forms of life on Earth] to see this kind of creature.

So we were on the shores of lakes that are very much analogous to what we are seeing on Mars, and, all of a sudden, we find those primitive terrestrial organisms just colonizing everywhere they can. So that's the cherry on top of the cake. We were not expecting that. We are hoping to develop this aspect, because, obviously, this is going to tell us a lot about the potential of these kinds of lakes.

Laguna Verde is three times saltier than seawater. I learned this the hard way when I tried to dive in it. Eleven pounds of lead would take me to the bottom without any problem. It was an interesting experience, an interesting dive. So they are stromatolite-like colonies. Because what I didn't mention is that we have all these fossil stromatolites that colonized the paleo shore of this lake. But, even better, there are still stromatolites forming today. And this is a very rare occurrence on Earth these days. We are walking on them—literally on the floor, and there is better news than that. I just received three weeks ago some analysis from the lab, and this is a dead ringer for [an existing species], the chemistry is the same exactly.

Our goal was the top of the volcano. And not only do you want to dive on top of this volcano, but you have to get there. We have to make sure that nobody is suffering from hypoxia, from any type of problem. We don't want to climb too fast, because we don't want to be anoxic at the summit. And, so, what we do is that we stop at night at 18,000 feet, at the only flat place we can find on that volcano. Basically, this is a 12 meter by 4 meter kind of flat, and we are jamming our four or five tents on that, with an 800-meter drop nearby. I didn't mention that to Code Q before we went there. But the slope is about a 40–45 degree slope, on very unstable material. We can't stabilize it, and the only thing we can hope for is that there will not be an earthquake that night.

So once we are there, we spend one night there for acclimatization, and once we are on top, the scenery is quite something. The lake itself is a receding lake now, about the size of a football field. And the paleo shore marks the level of the lake in the '50s. That also tells us something: those lakes are disappearing. It is really time for us to understand what is going on there, because they will be lost forever in a few years.

So when we went there, of course, we wanted to dive. We have several objectives when we are doing this. The first one is to characterize the habitat and

microorganisms and ecosystem that are there. And most of the summit lake, like Laguna Blanca or Laguna Verde, is a very shallow lake, which means that also the UV radiation—the microorganisms are going to be affected by that.

The other thing that we are doing is to do physiological research. At Ames, we have the astrobionics group, which is developing the "sea pods". These are monitors that keep track of the team. You are monitoring your blood oxygen saturation, you are monitoring your heart beat and apnea during sleep, because this is happening when you are in the mountains. But the thing we wanted to do here was a little different.

I am a free diver. I never dive with oxygen tanks, because I find them heavy and cumbersome. At 20,000 feet I had to convince many people that it was a good idea to hold my breath. But, basically, people in good health at this altitude will be at about 65 percent oxygen saturation in their blood. And their heart is pumping at around 135 beats per minute.

We wanted to understand what was going on with the oxygen saturation by doing the free diving. There are things we cannot do at sea level, because your blood is always saturated in oxygen at sea level. And if something happens to this saturation and heart beat system, you will see it at this altitude.

I also had another motivation—the very overarching objective is that, as an organism, your body is going to produce, one way or another, more oxygen, and we need to understand that. Because we have many people in this world who are dying from lung and heart diseases. And helping those people breathe better, or maybe find a solution that will help them, that is something that is really important to us. I took this point home because my father passed away from massive heart failure a few months before the expedition. And witnessing my father's last days in the condition he was in, and the difficulty he had breathing,

I said that the only risk I could take would be to have one more person to suffer that thing.

So, this is where, in my mind, there is no question about the risk entailed. If some people wonder why we are exploring, I tell you I know why I'm exploring.

So, we actually monitored the free dive. And we realized that people can be very well with a heart beat of 39 beats per minute—this is how low my heart went when I was diving there—and getting out at 20,000 feet of that kind of water, which is at 40 degrees, with a blood oxygen saturation at 93 percent. Something is producing oxygen in our body. Something remembers that we are aquatic mammals. We need to understand what it is. Nature gave me two good lungs (actually, Stanford is looking for the third one) and a good heart, and I want to take advantage of that to help make headway, to find solutions for one of the most horrendous killers in our society.

So, this is one of the aspects of why it is worth taking risks. On the other hand, on the NASA-related objective, these physiological monitors tell you everything that you want to know about a crew's health status. It was really

IT IS HARD TO THINK THAT EXPLORATION CAN BE DONE WITHOUT TAKING RISK . . .

THAT, ULTIMATELY, EXPLORATION CAN BE DONE WITHOUT LOSING PEOPLE ONE WAY

OR ANOTHER. EVEN IF THIS HAPPENS, WE HAVE TO MAKE SURE THAT WE HAVE DONE

EVERYTHING POSSIBLE TO MITIGATE THE RISK AND TO PREVENT IT FROM HAPPENING.

important for us to show that this was working, and we actually did some live transmission of our vitals directly to Ames and Stanford while we were there. That was the first time this was done. That was one of the co-aspects. The other aspect, going back to astrobiology and microorganisms, is that we are actually now in a position to know those microenvironments a lot better.

In the past two years, we have also been involved in pinpointing some interesting effects of the UV radiation. For example, diatoms are experiencing mutation and malformation due to UV radiation. We want to understand this because it tells us something. There are, apparently, mutations that we need to understand better, if it is a mutation toward extinction or if it is a mutation toward evolution. It is really important in a world where you have high UV impact, such as Mars would have been in the past. The other aspect of it is that we are living in a world where UV impact is becoming more and more of an issue. We have trouble with the ozone layer, and if these little creatures over there are developing sunscreen and a UV-resistance strategy, we need to learn about it, because that will help the general public.

Sometimes we have interesting encounters. One of a diatoms [planktonic unicellular algae] we found was thought to be extinct a long time ago. Well, it's thriving in Laguna Verde, thank you. It is having lots of success. It is still there. In another case, we found a diatom that was known to exist only in the Baltic Sea. Don't ask me how it ended up here, but it is there and thriving. Maybe the conditions of salinity and cold are very close to the Baltic Sea.

So we are learning a lot. Stromatolites were the first creatures on Earth, and they are still there. They are evolving and they are telling us about our past. If we are able to decipher what they are telling us, we will know better about our origins and, maybe, possible origins of life on other planets, because those lakes are very similar to that of Mars.

To go back to what many of the previous speakers talked about, managing risk and people, there is, obviously, a physical responsibility. You have catastrophic risk, of course. There is nothing we can do about the volcano exploding on us if we are there when this happens. Everything else we can manage.

I would say that the state of mind I would go with when I leave the U.S. and go to Bolivia would be to say that, as much as I can prevent it, nobody will get hurt on my watch. I am always working on this, and I have to manage a very diverse group of people. This group of people ranges from young people, students, to people who are senior scientists to people who speak different languages.

Last year in my team I had Hungarians. I had Chileans and Bolivians. I am French. I had American people and Spaniards. That makes for good jokes when we are trying to translate one thing to another. We have to make sure that the safety things will get through all the time. My personal standpoint here is that risk is definitely inherent to exploration, and it is necessary for discovery. And I will go a step farther. I say this is also the essence of survival. If we want to do it, we need to explore. We need to explore new ground.

I guess we have to share an ultimate responsibility, as Dale and Penny were mentioning. You have a team with you. You are responsible for them, and you have to have answers. The main thing is that you have to be accountable and responsible for all that. It is hard to think that exploration can be done without taking risk. It is hard for me to think that, ultimately, exploration can be done without losing people one way or another. Even if this happens, we have to make sure that we have done everything possible to mitigate the risk and to prevent it from happening, as much as possible.

Deep/Underwater Cave Environments

I've come from a rather unusual background in that I came up through engineering school wanting to be an astronaut, but had the good fortune of discovering that there were remaining places to be explored here on Earth in the meantime. As a teenager, I watched a couple of these guys who talked here this morning walk on the Moon. They were my heroes, and it was the U.S. space program that was directly responsible for my going out and getting a Ph.D. in engineering and wanting to work in space. In the process of trying to get into the Astronaut Corps at various times, I have also had the privilege of being involved with a large number of expeditionary projects dealing with things that go down into the Earth as opposed to things that go up. I added it up a little while ago. Over the last 26 years, I've spent 7 1/2 years in the field on expeditions, of which 353 days were below 1,000 meters deep underground, based from subterranean camps. So, I'm either a troglodyte or somebody who's looking for planetary exploration and hasn't been able to get off this pile of rock yet.

What I am going to do here this afternoon is to rapidly take you to three of the most remote places that humans have ever reached inside this planet. This is serious business. It is more serious, in my opinion, than high altitude mountaineering, because of the

William C. Stone
Engineer/Deep Cave Explorer and President, Stone Aerospace

William Stone led the exploration of Sistema Cheve (-1484 meters), (-1475 meters), and Cueva Charco (-1286 meters), among many other deep caves around the world. These Mexican caves represent the three deepest systems in the Western Hemisphere and currently the 8th, 9th, and 21st deepest natural abysses in the world respectively. During the past 33 years, Bill has organized and led 47 expeditions and has spent seven years in the field on exploration projects. He holds 11 patents and patents-pending and is the inventor of numerous other exploration-related tools. He is presently leading the three-year DEPTHX project for NASA to develop and field-test a prototype robot for the postulated Europa lander third stage, the "hydrobot" that autonomously maps the subsurface ocean of Europa.

multidisciplinary nature and the remoteness. I don't consider expeditionary deep caving as something you do for excitement. You do it because it's an opportunity to explore one of the last true frontiers on this planet. The classic distinction on this subject came from arctic explorer Vilhjalmar Steffanson, who once spent five years working solo north of the Arctic Circle. Interviewed about this one time, the reporter asked [Steffanson], "Well, you're an adventurer, aren't you?" He said, "Son, adventure is what happens when exploration goes wrong."

I have had that motto emblazoned upon my heart in letters of gold ever since. You do not get Brownie points for having your name on a tombstone. You have to come back. With that in mind, I have actually taken a lot of cues from how NASA trains its astronauts when preparing for, and staffing, expeditions. In the subterranean world, where we are about to go, it is a gloves-off environment. The exploration front is now getting to the stage where it is so remote and so difficult to reach that no matter what technology we have at our disposal, and no matter how Olympically-trained and fit the people are who are involved with it, we still get stopped. Every time you go for four or five months in the field, if

INTERVIEWED ABOUT THIS ONE TIME, THE REPORTER ASKED [STEFFANSON], "WELL, YOU'RE AN ADVENTURER, AREN'T YOU?" HE SAID, "SON, ADVENTURE IS WHAT HAPPENS WHEN EXPLORATION GOES WRONG."

you're lucky, you're a kilometer or two deeper into the planet. I am going to try to give you an idea here just what this world is like. I'm going to show here what would be the equivalent of summitting Everest and K2, but it's all going to be in one continuous trip proceeding down, in order to give you a sequential feel for the logistics and remoteness.

Rising out of the southeastern area of southern Mexico is the Huautla plateau. It jumps straight up about 2,100 meters. The top of it is cratered with gigantic sinkholes. The water that rains on this area for 500 square kilometers all goes internally and, in the process of doing that, it creates some pretty substantially-sized voids.

About 50 kilometers away to the south and a kilometer higher in elevation is Cueva Cheve. This was only discovered in 1986. This underscores the still unknown extent of this last frontier: Cueva Cheve was not even known to the modern world until 1986—just 18 years ago. The endeavors we're talking about require a lot of technology, starting off with the fact that you're going down. You don't just walk down that depth. There are three kilometers of specialized rope that are used to rig over 107 drops and traverses going down into this cave.

Typical vertical drops on some of these things can reach distances of 160 meters straight down. That would be about one and a half times the height of the Vehicle Assembly Building, for those who are here from Kennedy Space Center. A lot of times they're a little bit shorter, but they almost always have water accompanying them. The deeper you go, the more water you collect. The tributaries each add a little bit in until, finally, you're dealing with quite a serious flow. You're rigging lines on the walls to keep out of that, usually. This is life on rope. You're usually carrying around a 25-kilogram pack, which has the supplies that you're bringing. This is very much an inverse variant to the siege tactics that you see used on high altitude mountains where you're building Camp 1, Camp 2, Camp 3, and you start off with a pyramid of 50 people. Sometimes, we've had as many as 150, but 50 is a typical number for a trip for four to five months. That works out to be a good number to work with—you have depth of personnel to handle emergencies in situ. We've generally considered that a team of 12 to 18 is about the minimum you would want for safety and skill redundancy. And the further down you go, you're adding more tributaries until, finally, you're into rivers by the time you get to about the -800-meter level. This would be a little over a half a mile deep in English units. I know nobody around here works in English units . . . except when you're designing planetary craft going to Mars. [Laughter]

One of the things that we've had to deal with is the fact that we are constantly running up against the limits of human endurance. Typically, if you go on a marathon exploration trip, you can stay up for 24, 28, maybe 30 hours, and, after that, if you come back to base camp on the surface, you're out of commission for two to three days while you're recovering. You can't do that underground, because even when you are resting—say at an underground camp—you are consuming supplies, which are finite and paid for at great price of effort. If you want to move efficiently, your people have to be roughly 16 hours from anywhere that they have to go. And, so, we begin to establish a series of camps. I want to dispel the notion that a lot of people have that these places are claustrophobic. In reality, many chambers and tunnels in southern Mexico—owing to the significant tropical yearly rainfall—are so big you can't even see the ceilings or the walls. It really is like being on the dark side of the Moon. These are team endeavors. None of your specialized personnel— divers, climbers, surveyors—get to the "front" without riding on the sweat of a substantial support team. Like high altitude mountaineers, cavers frequently refer to such support crew staff as "sherpas."

Again, when I think of an expedition—and this is a sticking point with me and it may be with others—there are a couple of holy words in the vocabulary of true explorers. And one of those is the term "expedition." To me, this is an endeavor of 20 or more people being out in the field for four-plus months. That's a serious distinction. Anything short of that is what I would refer to as a recon mission. And, so, in the case I'm discussing here, you're on site for four or five months, and people are working daily. It is not uncommon during the early stages of an expedition for the lead rigging team to get caught between known camps, and, so, setting a bivouac is something you plan for as you move in. Generally, it is for logistics reasons that you don't get as far as you would like—for example, you might not have been able to carry enough rope to continue rigging.

Well, the reinforcement crew comes in the next day with another kilometer of rope and off you go. And then that support crew retreats to the previous camp behind them and the rigging team presses onward. And so it goes on down, until you are now roughly four days traveling distance from the nearest entrance. So, when you get to a place like this, you begin to think about the fact that you're pretty remote, and there really isn't going to be any rescue, except from the people who are with you, particularly if it's something that requires urgent response. If you can get a person back to a camp and stabilize a broken leg or something like that, you can always send out for assistance. But there are places where even that is not possible, particularly when working beyond flooded, underwater tunnels.

The most remote place that we've been so far is at the current limit of exploration in Cueva Cheve. When you get down to a distance of approximately 8 1/2 kilometers inside and at a depth of -1,360 meters, you're moving through river canyons, and you keep thinking that, well, this is just going to keep going down like this. But the problem is that caves are always quirky in terms of geology. All it takes is a slight counterfold in the limestone strata, and that river that was boiling is now static, placid, and leading you into tunnels that are completely waterfilled to the roof for substantial distances, before they usually rise back up into air-filled sections of cave. They used to refer to these places as terminal siphons. In fact, that's the name of a rock band I'm in, so you'll hear us touring someday.

So, you get down to a place like this and here you are, you're roughly 1,360 meters vertically down and 8 kilometers in. Everything down here is paid for preciously by the people who transported this down. You're living on ropes for days to get this here, so you have to be very careful about what you bring. At this point, those in the lead—in this case, trained cave divers—have this enormous pressure on them to perform. And I'm sure that everybody who's flown on a rocket knows this same feeling. Here's 50 or 100 people who have given of their time, of their lives, of their sweat for four months. That's not counting, by the way, in the case of many of these, that we have spent two to four months rehearsing with those same teams over the preceding one or two years. When you get to these places where the tunnels are full of water, now you're into another level of discipline where you have to be aware of the fact that, number one, you're going to be using portable life equipment—what astronauts refer to as PLSS [Portable Life Support System] units for EVA [extravehicular activity]. And that's the way I think of this. I think of this as EVA. And when you do that, you have to be thinking a couple of things. Number one is that anything can go wrong at any time, and so the best way to deal with this is to believe that this place is actively out to get you. When you think that way, you start making checklists ahead of time. In fact, we have them all laminated on waterproof paper. Not only before you go in, but after you come back out. It's the equivalent of pre-flight and postflight checklists.

But there's more to this. I'm going to say one thing here, and then I'll come back to another side of it. We've been talking about issues for reducing risk, one of them being making our equipment "bombproof." That's a bit tough when you look at the places that you have to go through to get these items or the equipment that these people have down here. And the other issue is redundancy. But before I get to that, there's this whole issue of peer pressure that is on you. You have to be ready at any second on one of these things that, if you get 50 feet inside this underwater tunnel and you don't like how you feel, you abort. That's a cardinal rule.

Somebody was asking me recently how many people do I know on expeditions that have died? And I really never thought much about it until I started adding it up. Over the last 18 years, I've lost 16 good friends. People that I've climbed with, worked with on expeditions, people who were very qualified.

SOMETIMES WHEN YOU GET TO THESE LOCATIONS, THERE IS NO DRY LAND, AND SO YOU'RE LIVING IN HAMMOCKS STRUNG FROM ROCK BOLTS ABOVE THE WATER AND HOPING THAT YOU'RE HIGH ENOUGH THAT, IN CASE THERE'S A FLOOD, YOU'RE NOT GOING TO GET WASHED AWAY. THAT'S THE WORLD YOU LIVE IN WHEN YOU'RE DOWN THERE.

The reason they're not here right now is because they went a little too far—they didn't abort when they should have, they didn't stop and say, "Wait a minute. There's a stack of things that are going wrong here." Nothing ever happens in one blow. Jim Lovell pointed that out this morning. But a string of little events occurs—you start going down there and you get tired. You get tired and you say, "Ah, I don't need to have this extra little piece of safety line here." And, "Oh, well, I don't need to check this other piece of equipment I have for descending a rope." Pretty soon, things start to add up, and you don't have that safety on there and when you sit down, one of those carabiners is unlocked and it comes unclipped, and there you are with 25 kilos hanging below you, and you're hanging on the rope by one hand. You know, that kind of stuff happens. So, you have to get religious in your discipline about how you deal with the technology. This is high technology exploration that's going on down here. More serious than a typical EVA mission outside the Shuttle.

The Portable Life Support System we use is all fully-closed-cycle equipment. There is no air in this system. It's all helium-oxygen running at about 6,000-psi in carbon-carbon tanks. The range on this particular rig is about eight hours and depth-independent. We developed this particular PLSS because we didn't know how far we were going to have to explore underwater. Sometimes when you get to these locations, there is no dry land, and so you're living in hammocks strung from rock bolts above the water and hoping that you're high enough that, in case there's a flood, you're not going to get washed away. That's the world you live in when you're down there.

In the bottom of this system you're doing 600 meters—that's roughly a half a mile—of diving, starting at the -1360 meter level. When we got to the other side, we established a camp. We were over there for a week and explored another three-and-a-half kilometers beyond the underwater section. These places represent the frontier right now. No one has been able to organize an expedition since to go further. In the case of Huautla, it has been 10 years since anyone visited that location; in Cheve, it was 2003, and it's unlikely that a return there will be fielded before 2007, although we do have that one in the planning stages already. Neither of these places are finished. They are beckoning. They are wide open. And, yet, we can't get the tactical logistics together or the team, because those people all have to be not only vertical trained, they have to be comfortable with where they are in a remote environment, and they have to be able to dive at the same time as well as climb. You start putting all these things together, all these various task loadings, and, pretty soon, the filters get pretty serious. There are not many people in the world who have all these skills.

This is the kind of thing that we are going to have to think about. Who are you going to put on the Moon? What kind of expeditionary-qualified people are you going to put on the Moon? Are you going to live together for a year up there? I mean, the psychological aspects are one thing, but the multidisciplinary training is going to be phenomenal.

If you get away from this logistical problem that we have of getting stuff down to -1,500 meters underground, and you can bring things in on tractor trailers, the whole ball game changes. One of the great natural wonders of the world, just south of Tallahassee, Florida, is a place called Wakulla Springs. Up until 1987, it was basically unexplored. You could look at the entrance from glass-bottomed boats, but nobody had seen much more than 100 meters inside on scuba. And we had an opportunity to go there with National Geographic in '87. But it was in '99 that things really got high tech. This project was a good example of how exploration need drove equipment development. Many pieces of technology we used did not exist prior to 1999. We built all this—a floating, saturation diving, recovery system; 20-kilometer-range propulsion vehicles; dual-redundant, closed-cycle PLSS backpacks; and 3-D, automated mapping systems—on a two-year schedule, once we got the go-ahead from National Geographic and corporate patrons. We had a dual closed-cycle PLSS; six onboard computers running this thing; twin head-up displays; 18-hours range. There is no EVA system in use by NASA that would meet this standard, and I'll tell you why. When we get down there, we're going to be doing missions to distances of about 4 1/2 kilometers from the entrance at 100 meters underwater depth. When you're out there, you are two and a half hours one-way travel time from the nearest egress point. By comparison, if you're on a Shuttle mission or a station mission right now, and you have a problem with your suit, whether it's an Orlan or an ILC Dover suit, you've got roughly 10 to 15 minutes, maybe 30 minutes in open circuit mode at best, to get back to the hatch. Here you don't have that luxury. You are two and a half hours out, and you've got to figure out how you can do it.

It's important to point out that everybody here was in a situation where, no matter where they were on that trajectory, they had an abort scenario. The maximum mission duration by pre-agreement among the teams was 10 kilometers penetration. And you had a factor of four to one safety margin to get yourself back. Everybody who was doing these would rehearse each mission beforehand for upwards of 12 hours (in the spring basin), and have people come in and tell them, for example, that all of their vehicles were dead, they had to transfer and get towed out by one of their partners. And then you'd tell them, for example, that half of their life support system was gone and they'd have to continue out on that. So you're continuously rehearsing all the various abort scenarios. In that sense, it's no different than training for a space mission. But here, we explicitly included abort capability for every phase of the mission. That isn't true of space flight currently.

A typical mission is all about collection of information about what the frontier is. If you're an explorer, you should be out there trying to collect that data, just like Steffanson said, and get back safely with it.

We built for this project a gadget—known as the "Digital Wall Mapper"— with a very high-grade, inertial guidance unit, phased array sonar, and about eight computers. And it's imaging the wall as you drive through, to build a three-dimensional map of the aquifer. The person in back is the safety diver. Their job is to make sure that when this person is driving this monstrosity, that if they do something wrong, they're going to come up and help them out.

Typically, we would have a support crew follow people in to a depth of about 80 meters. Perhaps 150 meters into the cave the crew is running on helium-oxygen, and they'll have 18 hours supply down there. The people on top have got the biggest sport diving cylinders in the business, and they're dumping out about a thousand dollars worth of helium-oxygen on that support mission just because of the fact that they're wasting it away, breathing it into the water column.

And then off you go for approximately the next five and a half hours, with a typical run down here to 100 meters depth, and then you're back. Now you can't come straight out, as somebody like Mike Gernhardt or anybody will tell you who has done EVA. You have a decompression issue, a very serious one. In fact when you're five hours down at 100 meters, you're almost saturated. At 60 meters down, we had a string of closed circuit cameras following people out. They had flashcards so they could tell if everything was OK, or they needed supplies from down there.

When they reached a certain stage—the completion of their 30-meter decompression stop—six support divers would drop out of the sky, like something

75

out of a James Bond movie, and come down and take away their closed-cycle unit, give them a regulator, and let them switch up into a pressurized, personnel transfer capsule. They would then be transferred up to the top and into what we call a mini-saturation system, where they would spend the next 12 hours. Total mission duration, about 22 to 24 hours. We did this everyday for three days, took a day off for maintenance, and then did it again, for three months while we were down there. No issues with safety on that entire project with any of the mission crews.

Ten million data points were gathered on this trip to build the world's first three-dimensional cave map. So, in our case it's not only a desire to be out on the frontier, but it's also about bringing the data back home.

A Writer/Climber Perspective

I've been freelance writing for 25 years, mostly about adventure, exploration, spinoffs to archeology and history. But before that, I was a serious climber for about 20 years, from age 17 to 37. I didn't climb high altitude peaks like Ed, but specialized in technically difficult peaks in Alaska. And I still climb, although pretty pitifully.

But for me, the most critical question of my life had to do with the fact that by the age of 22, I had been a firsthand witness to three fatal accidents to partners. It began with my first partner in high school, Dave Lee, who, only four months after we started climbing, was killed on the first Flatiron above Boulder. It was not a very hard route, but when we got a rope snagged and had to unrope, he climbed down solo to retrieve it. He slipped and fell 700 feet, with me watching.

Three years later, I was the first person on the scene when two guys fell out of Pinnacle Gully in Mt. Washington, one of whom was actually in training to be an astronaut. We tried to resuscitate them to no avail, and then had to haul their bodies down to Pinkham Notch. And then, just three months after that, on my hardest Alaskan expedition to Mt. Huntington, after all four of us had reached the summit, on the descent, Ed Burns, the youngest of the four of us, a 20-year-old sophomore in college, inexplicably fell, had a

David Roberts
Writer

David Roberts is a freelance writer living in Cambridge, Massachusetts, who specializes in adventure, archaeology, and history. He has co-led 13 mountaineering expeditions to Alaska and the Yukon, making many first ascents, including the direct north face of Denali, the west face of Mount Huntington, and the southeast face of Mount Dickey. He is the author of 16 books, most recently *The Pueblo Revolt: The Secret Rebellion That Drove the Spanish Out of the Southwest*, published this month by Simon & Schuster in October 2004.

rappel come loose. And we were descending in the middle of the night, we could just barely see in the Alaska twilight. And without my knowing, even today, what happened, the rappel fell, and he simply fell away from me 4,000 feet, and we never could even look for his body.

The question to me now is, why did I keep climbing after these terrible experiences? At the time, I didn't even really examine it. I mean, I came close to quitting, but I didn't really examine why I kept climbing. I think if somebody had asked me, I would have said, "Well, wouldn't their deaths be even more pointless if I quit climbing?"

So, when I tailed off climbing and got interested in writing about other climbers, including Ed, I got much more interested in this whole question of risk, motivation, and the risk-reward payoff. And, in 1980, I wrote an essay called "Moments of Doubt," an apologia for doing something as crazy and risky and useless as climbing.

My favorite climbing autobiography is Lionel Terray's *Conquistadors of the Useless*, because it is an awkward, but telling, phrase. And just this year, I've finished a memoir about my climbing in which I actually come to the opposite conclusion, that maybe it wasn't worth the risk. And what changed my thinking about this was that I had always thought about risk in terms of the question, basically sort of a solipsistic, selfish question, does the reward I get for making a first ascent make up for the risk and the tragedy of someone dying? And it took me 35 years to realize that that was just a completely self-centered and, therefore, sort of stupid question.

And one of the things I did a few years ago was to go back and recontact Dave's brother and sister, the only surviving members of his family. Forty years

after the accident, and I had never communicated with them. And I found that his sister, in particular, was still in a rage with me over that accident, and that she lived with it every day of her life. We spent the most intense seven hours straight in Seattle on a park bench talking about it, and trying to untangle her feelings about it. And I finally realized that the question of, is it worth the risk, is not one you can really just apply to yourself. It really does involve family, the larger society, and humankind, ultimately.

I guess if I still had to come up with a rationale for climbing mountains, it would not be anything to do with the thrill of it versus the penalties, but, rather, whether the very endeavor has something inspirational about it that lasts. And I think by analogy with Scott and Shackleton and Amundsen in the Antarctic, they justified their expeditions in the name of science, just as I think NASA is continually doing. And Scott, when he died, had 30 pounds of rocks on his flesh, geological samples. He tossed away everything extraneous, but he somehow thought the 30 pounds of rocks were worth bringing back. Apsley

Cherry-Gerrard's worst journey in the world was to gather penguin eggs in the middle of winter from Cape Crozier, and under the misapprehension of the now exploded theory that ontology recapitulates phylogeny, penguins being the most primitive birds in the world, their embryos would tell us all about early human evolution. It's complete bunk. He performed the worst journey in the world to gather three penguin eggs. But we don't remember Scott, Shackleton, and Amundsen for their science, we remember for their example of daring adventure and exploration and going where no one else had ever been. To me that was the first, and is still the most exciting, part of exploration, to go where no one else had ever been. The only rationale for it is if it inspires other people. Ed Viesturs clearly inspires other people. I have been at talks he has given where the groupies have just gathered all around him. They want more than autographs. Is he a rock

I GUESS IF I STILL HAD TO COME UP WITH A RATIONALE FOR CLIMBING MOUNTAINS, IT WOULD NOT BE ANYTHING TO DO WITH THE THRILL OF IT VERSUS THE PENALTIES, BUT, RATHER, WHETHER THE VERY ENDEAVOR HAS SOMETHING INSPIRATIONAL ABOUT IT THAT LASTS.

star celebrity? No, but there clearly is something that touches the human spirit and, as corny as Bush's line is, "the desire written in the human heart" really does have to do with risk and exploration.

One last note, a curiosity from the Renaissance. We're taught, on and off, about the Renaissance, but someone calculated that on a typical Renaissance voyage to the New World, sailors stood a one in three chance of not surviving. And that makes Everest look like a piece of cake. It makes Bill Stone's stuff even look safer. And the collectors of the narrative of the voyages write often about the adventurers, but the adventurers were not the sailors, they were the guys that put up the money. The real adventure was taken by the financiers who backed the expedition. The sailors were just expendable work hands. All of this needs to be put in a historical and multicultural perspective.

Discussion

CHRIS MCKAY: Okay, what I'd like to do now is to go to open discussion. Keep in mind what you heard, and keep in mind the questions that would be relevant to NASA, in terms of sending humans to explore. Points that David made—is it pointless to send humans, especially now with technology advancing and, people argue, we could send machines. And it ties to these questions that David raised. And what I would like to do is have some audience participation and discussion. You can ask questions of anyone you want, or you can make a short statement.

QUESTION: Darlene Lim, NASA Ames Research Center, and my question is for Bill Stone. I wanted to find out what some of the advantages were that you presented, to say it would be more advantageous to send humans into the aquifer and do the mapping that you outline versus sending in submersibles?

BILL STONE: That's a fantastic setup. There might be one or two others in this audience with similar background, but I think I've found myself in a unique position in having spent about a decade of my life designing spacecraft and another decade designing robots, and now robotic spacecraft. During "CFT," or "copious free time," we do some work underground. The answer to your question goes like this, and I'll try to put it to you as succinctly as I can. There are places where robots are entirely logical to be used first, and I classify those in two situations. First, places that are lethal to humans, and second, places which are currently too remote to get to send people to with our primitive propulsion technology systems. So you send robots in those situations.

Everything else, in my opinion, is best done with people. Now, at some point, you do have to draw the line in the sand to say this is too risky. As I said when I was looking over that list of departed friends—no different from Jim Lovell citing off a list of test pilots he has known—these are dangerous environments. The reason we are still here is because we knew where to draw the line in the sand. Even if we do that, there is still unquantifiable risk, so the game is to say, all right, you want to be out there in person. Probably everybody in this room wants to be out there in person. Do I want to be here today? To be honest, I'd rather be at Shackleton Crater on the Moon right now, for a couple of years with a dozen good, qualified people. I'd be happier there than I am here or at work. Being on the frontier is what I live for. That's where people like Ed Viesturs live. If you don't understand that, you need to go on an expedition somewhere remote. Maybe you'll like it, maybe you won't, but among the people I work with, you can tell right off the ones that are concerned about where they are . . . you see it in their eyes when they get to someplace like Camp 3 or Camp 4.

In fact, people that I've worked with for years on expeditions have a name for it. They stole it from Jacques Cousteau: it's called "rapture of the deep." You can tell because their eyes are getting wider, and they're sitting there at those camps going, "I need to be concerned about getting out of here," rather than focusing on the job at hand, which is pushing the exploration frontier. The people you find that have a bigger smile on their face the deeper you go, those are the people you want, those are the ones you want out there with you at the frontier. Now, to go back to your question directly, why do this in person? Because it is the most stimulating thing that you will ever experience in your life. I have had the privilege of being where no one has been before many times. It is difficult to put into words to those who have not experienced it just how moving that feeling is.

If I had to capture a 20-second image on film that depicted that event and what its importance is personally to anybody who does it, it would be [from] the film Apollo 13 that Jim Lovell made with Ron Howard, where they show the image of flying by the Moon and the astronauts thinking, "I could be down there, lifting that rock up and looking at it in my hand." That's what drives us on. If you look at what we do in our normal lives, it pales in comparison.

CHRIS MCKAY: Bill just made a very good case for human exploration. I think we want to hear a countercase, so somebody in the audience think of a good case for why the space program should just be robotic, why we shouldn't risk human life. If there's no one in the audience who can state that case, we don't have a well-selected audience. Somebody come up with the counterpoint, the argument (so we can pillory them) why robots are sufficient. With that enticement, we'll go to the question here.

STEVE COOK: Steve Cook, NASA Marshall. I won't take that question, but I will say, first off, I think there's a gold mine of analogies here that are applicable to exploration that we need to capitalize upon as a country, to sell and sustain a vision that I don't think we're doing as well as we could today. With all of the experiences you've been having, a lot of these are firsts with me and I've been with NASA for several years. Question now with risk, with respect to Ed Viesturs and to Bill Stone. Ed, you talked about making the decision to hold back—you had the camera crew, this would have been a perfect time to go; your friends did not. I'd like you to talk a little bit about the criteria you used versus what you think or know that they used. Why did they go forward versus why did you stay?

ED VIESTURS: When we go to Everest, we spend six or eight weeks preparing for that final day. That's carrying loads, building camps, acclimatizing. And then the idea is that sometime in May, historically, we know that there's going to be a window of favorable weather that would allow us to do that final 3,000-foot climb. So the whole preparation leads to that. Once you're ready, then you have to take into account everything that you see around you. The weather patterns—now we can get good weather forecasting. In those days, it was just kind of starting to work, but we would have to base our judgment and what we were thinking about doing on what we were seeing, what we were feeling; our gut instinct has a big role in how I make my decision—you know, what am I seeing and how has that played a role in what I've done in the past. And we made a group decision—David Breashears, who'd climbed Everest twice, I'd climbed it four times by then— as a group, we all felt that the weather patterns that we were seeing were not historically what we were waiting to go to the summit with. So we made the call based on that and, also, just the fact that it didn't feel right. And I think Bill mentioned that, I think the reason a lot of us are alive today is because we know when those red flags are popping up, that we have to listen to our instincts. If we discover later that we made the wrong decision and the weather was good, well, big deal. We erred on the side of safety and being conservative.

Our friends, they had this date in their mind, and that was May 10th. And, a lot of times, climbers do that. Come hell or high water, they will just go for the summit on that particular date—because it's auspicious or whatever. And I've always said, ultimately, the mountain decides when you can go up, and you have to listen to what the mountain is telling you. There are signals, and you can't blindly, stubbornly go up, because we all know that Mother Nature is much stronger than we are. They did start on a good day—it was perfect. And had they turned around in enough time, they would have escaped the storm, which came late in the day, which is something that we saw developing every single day the past two weeks previously. But, again, they got so close to the top, they pushed further and further and further away. That umbilical cord of safety got stretched and finally broke. And I can't second-guess the decisions that they made. I wasn't up there. I can evaluate what they did. I can say that I wouldn't have made decisions the way they made them, but, again, as Bill said, there wasn't one decision that was wrong, it was a multitude of little problems, and then the straw that broke the camel's back was the storm, and then people died.

CHRIS MCKAY: Ed, I'd like to ask you a question. Do you have a personal short answer to the question of why you climb mountains?

ED VIESTURS: I'm a stubborn person, I like challenging projects and, for me, climbing is the most difficult physical and mental thing I can think of doing— especially without oxygen. And I know I have to train, prepare, think about what I'm doing, to be successful at it. And there's a huge struggle to get to those altitudes without oxygen. And so many things can go wrong, and when you succeed, and when you finally are standing on the summit of Everest, it's an amazing feeling. And it's something I can't find anywhere else, and, so, I've become addicted to it.

CHRIS MCKAY: Would you classify what you do as exploration?

ED VIESTURS: I explore myself. And I think that's the interesting part. I think, also, you can be in the middle of the desert, you can be down in the ocean, if you get in trouble, somebody can come and get you. Up there, you're on your own. If you make a mistake, you have to get yourself out of it. And it's a very rare feeling. As normal, sea-level, landlubbing people, we don't get that feeling of isolation where we think about every single step and every single move that we make, [that] there can be consequences because of those. And that's an interesting thing that I feel up there.

CHRIS MCKAY: Thanks. We have a question here.

QUESTION: Dennis Wingo, Skycorp, Incorporated. And this is for Neil, Ed, and you, Chris. I see a commonality between what you do on the mountain, what you do underground, and what you do in the Arctic. And the commonality is a staged approach, abort modes—you don't do everything in one big, fell swoop. And it seems that in the space arena that we try to do that, whether it's a manned system or whether it's un-manned. But especially in the manned, because there's a religious argument (I call it) going on now at NASA: Do we build a heavy lift launch vehicle, or do we use the assets that we have to go back to the Moon and on to Mars? It seems that with the assets that we have now, and that we [will] have in the near term, we could use your staged approach to go to the Moon and go to Mars.

ED VIESTURS: Yeah, for us staging is critical, because we have to not only put supplies at various camps, but it's also the process of acclimatization. So, you need to take the time to go at various altitudes, and then come down to recuperate, and then, slowly, work to a higher altitude. It's kind of a necessity in both aspects, acclimatizing and, also, then, getting your gear in place. Once acclimatized, though, I've developed a system now where we'll go very quickly and—like you said, in that big launch vehicle we'll carry everything that we need, and, in three days we'll do what normally takes two months. We'll climb, moving everything with us, and then go to the summit. That's riskier, but it's also faster, because you're spending less time in a dangerous environment. So, I think the trade-off is worth it. I'd rather climb faster, without the series of camps behind me, but I need to do that initially to acclimatize.

CHRIS MCKAY: Bill, you want to give us the downward view of the same thing?

BILL STONE: It's no more risky than being a test pilot, in my opinion.

CHRIS MCKAY: And your point?

BILL STONE: No, I mean, you can control the risks. There's one thing that you can't control, but you can prepare for, and that's the weather. And we're going to have that kind of phenomenology wherever we go, whether it's a big storm that comes in on Everest or whether it's a hurricane that dumps ten inches of rain over the plateau and you get a 10-meter-high rise in the water wherever you're at. I have friends who have bolted their way 30 meters up a canyon wall in New Guinea as the water rose behind them. They hung from rock bolts for 32 hours waiting for it to go down. They were prepared, you know. Yes, they were in a dangerous place, but they had their act together.

Let me extend this to space, because I have actually put a lot of thought into this. I've designed space systems. I've designed reaction control thrusters. Things like that. One of the things that apparently has been misunderstood here is that we did have complete control over the situations where we were—even the weather. All of our camps were selected to be above the water. The way the ropes were rigged—all of that stuff was of very high reliability. Further, we trained everyone on the team to know when to replace things, if they discovered that a line was fraying, for example. And there were depots of spare rigging tackle situated at various locations along the way for just such emergencies. We even had ways to get out if a rope broke—equipment for scaling overhung domes. We had all these things covered.

Now, if you switch to trying to think about what are you going to do if you're trying to get to the Moon—which is the next logical target for us to get back to here, and prove that we can live off this planet—the first thing you've got to say is, all right, how are you going to get to low earth orbit (LEO)? And there are only so many ways that we can do that right now.

You've now seen what I do. If you were to say to me, "Would you fly the Shuttle?" or something like that, then I'd have to say, "I have to think very carefully about that," because it does not meet my criteria for having an abort mode at any point in the trajectory. You didn't pay me to come out here to give you sweet talk, so I'm going to talk to you straight. The last time we had a launch vehicle that met those criteria is when these [Apollo] gentlemen were flying. Why on Earth have we not done that now? That's just my personal impression. If I was going to try to put something up there, we'd probably be using ELVs with abort modes to get us out to LEO. Beyond that, you're into the issue of how survivable are your vehicles? You need to be thinking in terms of propulsion, life support, and everything else.

You guys are great at this. NASA is a technological gem on this planet. But, unfortunately, you don't have any true "expedicionarios" here, as we say down in Mexico—true expeditionary people who think, "Two years from now we're going to run an expedition to the Moon." And then go do it. The way NASA thinks these days is, "We can develop these technologies and, maybe, 15 to 20 years from now we'll be back on the Moon." That's too late. That's too late. You've got to get there now and learn from the frontier, just the way we've been doing underground here on Earth. Those places you've seen today were unexplored 12 years ago—completely unknown. We built the technology to go there. Given the enormous resources at NASA, there is no reason that we can't be back on the Moon within five to seven years, max.

QUESTIONER: If the Moon is where you want to go.

BILL STONE: There you go!

DALE ANDERSEN: I want to go back to your question about staging, which was doing things in steps. I think a lot of the issue there is knowing what the stages ought to be. I remember once getting a lot of flack—we were going to the Antarctic to do diving under the ice. I put in travel orders to go to Key West—that was our staging spot. We were going to do training for ice diving in Key West.

Now, maybe you could argue that was a good place to go to stage the Antarctic, or maybe you could argue that it wasn't necessary. So, when we plan anything, any expedition, I like to think: What is our goal? What are we really trying to do? What's the driver at the long end? And, then, what do we have to do to get there? If you look at the way Apollo worked, as I understand it, it was very much that way—as we heard this morning. You want to go to the Moon, so you need to develop docking, you need to demonstrate that you can stay in a spacecraft, and so on. Take something like Gemini—it was a requirements-driven program: If you want to go to the Moon, you're going to have to do this first. It wasn't: Let's do Gemini and see where it leads, or some nebulous concept that, somehow, if we develop these things, a mission to the Moon will miraculously appear from the pieces.

And, I think, as we think ahead to an exploration program, we have to do the same thing: Where do we really want to go, and, then, what are the pieces that lead us there? And I think that's also what Bill was saying. So your question is a very good one, and I think exploration on Earth, and people that plan expeditions on Earth—the logic that they approach can be applied, but the answer is not so easy to come by.

CHRIS MCKAY: Okay. Maybe someone can address my challenge to pick up the case for robots, but probably not you!

QUESTION: Well, maybe I will! I'm from Goddard Space Flight Center, a scientist-type as opposed to a program manager, which seems to have been commented upon today.

CHRIS MCKAY: Derisively, I might add!

QUESTION: It seems to me that over the years NASA has been very ambivalent about people in space and science. Back in the '60s, getting people up there seemed to be something done for its own sake. Putting somebody on the Moon was done for its own sake. Then, in the '70s and up to maybe the '90s, we started getting into a mode where we were trying to be sold on the fact that people had to be there to do the science better. And it never appeared to me that that was the case—that it was a selling job to justify getting people in space, for the most part.

And I guess my question now is: do you think that the public, Congress, and the media are ready to do real people in space as opposed to trying to do the science with robotics? My personal feeling is much of the science could be done better with robotics, but we're not going in that direction right now.

CHRIS MCKAY: Okay. That's a good start to the case for robots. And I think we need to put that case out there as part of the complete discussion, because one answer to the risks is, don't send people. You avoid the risks. If we go back to Dale's categorization of risks—mission risk, personal risk, and team risk—what you eliminate is the personal risk and team risk by not sending people. Certainly, when Steve Squyres did the MER mission, there was risk. But no one's life was at stake on Mars—it was just a programmatic risk, a science risk. The worst case that could happen was that Steve would have wasted seven years of his life and no publications would have come out of it. Steve might have felt that that was a disaster, but no one would have gotten killed over it.

So, I think the case where we think about risk—we can't just assume that means that humans are in the loop by definition and, therefore, we figure out how to deal with the risk with humans. We have to step back once and say: are humans even an essential part of the program, of the loop, or do we try to do science by robotics?

So, now we've made the case for robots. Let's let the panel and the audience react to that case. David is going to add to that. We've got a momentum going here!

DAVID ROBERTS: No, I actually would strongly believe that robots are the answer. And I think it's probably my father's influence, because he told me years ago, as an astronomer, that there's no way we're ever going to get very far in human terms in space, so the future is going to be robotic. And I found the Mars rover landings more gripping than the original Moon landings, human-equipped Moon landings. And I would say that the Hubble Telescope was far more important and exciting to me than any manned travel in space.

And I think that, for the first time in history, maybe we can actually make the emotional and psychological investment in machine discovery—as we're also doing in the deep sea—in lieu of the conquistador going out there and doing it himself.

CHRIS MCKAY: Okay. Penny, I'm hoping you'll take the contrary view!

PENNY BOSTON: You know I am!

CHRIS MCKAY: What a surprise!

PENNY BOSTON: I think that there is no dichotomy between robotic and human exploration. I think much ado is made out of that. They're obviously context-dependent. There are strong reasons why we're interested in human exploration beyond simply the scientific function. And I disagree that now, or any time in the near future, or even the mid-term future, we can design a robotic instrument that can have the capabilities of the tremendous flexibility that a field scientist can have.

However, that being said, I am a great fan of robotics missions. I love robots— I wish I had whole fleets of them myself. I'm trying to get MIT [Massachusetts Institute of Technology] and JPL [Jet Propulsion Laboratory] to build whole fleets of them—because they're a tool. They're not a viable life form at this point. At this point, it's not as if we're going to send robots or people. They are obviously complementary to the whole scientific process. But science is, fundamentally, a human enterprise, and the value of science is, fundamentally, to us as humans.

And, so, therefore, cutting us entirely out of the loop, I think, is inexcusable.

And, so, wherever in the solar system we can send people, where it makes sense and where people can significantly contribute—and one of the things that they have to contribute is that perception of the human experience of exploration—then I think that we should endeavor to do that, and save the robots to be our helpers and to go places where we can't go, to go first. And even on Earth, where we're exploring certain caves where we have tiny channels that we can't get into, even microrobotic devices there would greatly enrich our scientific exploration.

CHRIS MCKAY: Well said. Dale, and then Nathalie, can respond to the same question, and then we'll take a question here from the woman in black, and then Jim Garvin. Dale?

DALE ANDERSEN: Briefly, I was just going to actually completely agree. I think it's a mix; it's not an either/or case. When robotics are required they should do the job, and when people have the capability to go to those places, people should be in the loop.

I've used both robotics and going there myself, and I have to admit, I've been underwater with robotics, for example, while people at Ames have been diving with me virtually via that robotic device. That's a great way to share your experience in a remote location with a greater population. But it's not an either/or thing; it's just the right tool for the right place at the right time.

CHRIS MCKAY: Nathalie, could you add a little to that? And then we'll go back to general questions.

NATHALIE CABROL: I'm supporting both Dale's and Penny's views, because I've been putting together in the field an astronaut and a robot. And neither are always best, but they complement each other. And for exploring a planet, the human being will bring in his immediate background and an understanding of what's around him, which a robot cannot do now and will not be able to do in a long time. But, by the same token, the robot does not care too much about the environment. Is it cold, is it hot, is there lots of UV radiation? If there is an opportunity, they just don't care, they can last a long time at the surface. Together, they are almost an invincible team, but I would go a little step farther than that. I would say that, no matter what we think about it, exploration is within our genes. This is where part of evolution is right there in us. We wouldn't be here if we wouldn't have been taking risks and going from one place to another and exploring diverse habitats. And another planet is just the next frontier for us; there will be farther frontiers than that.

QUESTION: I'm Becky Ramsey, I'm from NASA Headquarters, and I have to say that regarding human versus robot, I have to come down on the human side, because I want to go. Yeah, and that actually leads into an issue I want to raise. I want to touch on something that we talked about earlier this morning, and that's individual versus government exploration. To bring up the example from this morning, Burt Rutan is ready to launch in just a couple of days here. What he's done is private financing, small group, hand-picked people. And that seems to be very similar to what you all do, whether it's individual or privately financed or government financed, even if it's a direct NASA project. What you do seems to be very small—one person, five people, even a hundred and fifty people—when you compare that to my colleague and I who work in the same building and have not met before today. That's very different. You know everybody on your team, and you're getting a lot further out there than we are. Is there something to be said for that—can we do this? Can government do this, can an agency like NASA do, in space, what you have managed to do here on Earth? Or do we need to find a different way and look at smaller, more team-focused models to do this?

CHRIS MCKAY: That's a really good question, and when we look beyond Earth orbit to distant destinations like Mars, the question becomes more pressing, because those teams have to be, by light-travel time requirements, more and more autonomous, and less dependent on remote control from Mission Control. Anybody want to address that, comment on that? Penny and Bill, think of an answer, too.

PENNY BOSTON: Okay, I'm thinking real hard, that's a really serious question people are wrestling with. The kinds of expeditionary things that we do are very small compared to a full-on mission to another planet. Therefore, the sheer number of people involved is so large in order to pull off a mission like that. In some ways, I'm not entirely sure that a lot of our experience in these smaller units is directly applicable. Because, by force, you have to involve so many more people, and the level of planning complexity far exceeds anything that any of us do.

So, the question is, can you do that in a governmental environment? Well, I think NASA is doing it in a governmental environment. And I don't see why, fundamentally, that transition from the kinds of missions that we're doing now can't be applied to also incorporate serious, meaningful, human exploration, plus an ongoing program. I don't see that it's not possible, with the caveat that, for certain applications, perhaps small companies are better, for certain limited things. Burt Rutan is also not doing NASA. He is not doing NASA in a can, basically. He is doing a very different scale of things than NASA has to worry about.

CHRIS MCKAY: Okay, I'm going to skip Bill's answer, because I think Penny hit the nail on the head, so we're going on to the next question, which is Jim Garvin.

QUESTION: Well, thanks Chris. I'm Jim Garvin, NASA Headquarters, Moon-Mars. I wanted to comment, and then address, an issue to the panel and everyone about the robot/human dichotomy, because I don't think it is one, and I think we have to pay attention to the great observations you all made. Because, in our history of space exploration, it was the robots that did the reconnaissance, the advance planning, and let the humans, like the great courageous heroes today, do the work. And I would submit to you that it was the humans back on Earth, and Apollo 17 and the others, continuing the work of the robotic spacecraft. So

89

it's that partnership that's important. And in many of the cases you've talked about, robotic reconnaissance wasn't needed to open the frontier. The humans could do that on the fly.

I would dare say I wouldn't want any of you brave people to go to the surface of Venus for the first time. It's probably better for our robot friends to do that. And today, as we think about Mars, I would submit that we're learning through the rovers that Steve [Squyres] will talk about. And what we're planning in the future is the reconnaissance necessary to go to those sweet spots on Mars. So then the humans become important onsite to do that kind of work. And I always marvel, if we think about this dichotomy, humans and robotics work together when we look at the samples brought back from the Moon from the Apollo mission. These multikilograms, each one itself a mini-universe for robotics and people to work together with here, to understand that world on the Moon. Imagine that anywhere we go, whether it be on Earth or beyond.

So, I look at it as the reconnaissance that's important, and today, a lot of that reconnaissance is better done on Mars, on the Moon, on Venus, way out where the origins of the universe are, by the machine. The question is, here on Earth, how can we amplify your experiences in these unique environments to better train us to use that reconnaissance to make the tactical decision to put humans on site, because we need to because it's in our gene pool, or it's necessary; it makes us better samplers. That's the question that I think this risk conference is treating, and I think, at times, that unfortunately comes down to the ugly words "programmatic cost." Where is the timing of that benefit?

CHRIS MCKAY: Okay, good comment. Let's go to Keith for another comment or a question.

QUESTION: Keith Cowing, NASA Watch.com: Okay, you were trying to pick a food fight a few minutes ago, and I love a food fight.

CHRIS MCKAY: I'm trying to claim my role as moderator to stir the conversation.

KEITH COWING: Well, in that case, I'll take that on.

CHRIS MCKAY: You can always fire the moderator, you know.

KEITH COWING: Whether it's robots or humans, in essence, it comes down to us going out there, whether we do it first or second or in tandem, and I guess one observation to make, to throw a little raw meat into this argument, is it's all about what we as a culture are looking to do. When is the last time somebody threw a ticker tape parade for a robot for doing something in space? When's the last time we all cried at a national funeral when a robot didn't work? Just an observation.

CHRIS MCKAY: Let's go to Donna Roberts, since I skipped her in the sequence. Sorry, Donna, you had your hand up a while ago, and then we'll come back to you, Steve.

QUESTION: I'm Donna Roberts, of the University of California San Francisco Medical Center, and my question is, with humans and extreme environments, where safety is utmost, and pushing physical capabilities, what is necessary for medicine? Should it be a first aid kit, should it be medicine, should doctors be there, what kind of diagnostic and therapeutic capabilities should we have?

CHRIS MCKAY: Can we get some quick answer? Penny, Bill, we'll let you answer this one.

BILL STONE: You just got me fired up. If you asked me to go plan an expedition to Shackleton Crater or something, who would you take? And I've thought about this, the answer is, number one, you bet, you're going to have a physician. He/she also would probably be cotrained in dentistry. You may have to have a back-up to that person, as well, who is perhaps cotrained in something else. We try to have a physician on every project that I have ever run. It's just too good a capability to have at base camp, whether they're out in front or not. It's one of those contingency things you have to think about when you're talking about long-duration projects. OK, if you're at a place where you're only an hour and a half away in orbit before you can drop back in—like you are right now at the space station—it's not such a critical thing, provided you have on-demand reentry. You're not going to have that on the Moon. So, yeah, you will have to have physicians out there, no question about it. The question is, what other skills should they have? And if I had to pick two, I would say emergency room experience and dentistry. The other surgical disciplines can be actually done through telesupport from the ground, as long as the individual on the expedition has basic surgery skills, as most ER types have. You can go on from there with all the other skills as well. My estimates for a sustainable lunar base exploratory mission run from 12 to 18 individuals, not unlike the minimum critical mass we currently use on deep caving expeditions.

CHRIS MCKAY: Isn't that something—you go to Mars so you can visit the dentist. That's why we need to send humans. Robots don't need dentists. We have a question here, and then we'll get to Steve. If you can pass the mike to Steve Squyres. A question here?

QUESTION: This is for Doctor Stone. Scott McGuinness, submariner, a student here [at the Naval Postgraduate School]. Doctor Stone, as a diver, I was looking at your videos, and you go deep into these caves, and you dive. U.S. Navy tables and the Haldanean model don't take that into consideration. So, my question is, how did you mitigate the risk of decompression illness (DCI), and how can that compare to the risk of radiation and all that for astronauts going to Mars? It seems like one of NASA's larger problems is how do we mitigate the radiation hazards of deep space.

BILL STONE: Decompression is not a big issue for lunar and Mars missions, at least from my standpoint. Mike Gernhardt and I could probably debate this for another 8 to 10 hours. The answer to your question is, those life support devices that you saw there had triple parallel redundant decompression engines running in real time—that is, the algorithms were running in parallel on three separate processors at the same time, such that you were guaranteed a valid tissue tension even if two of the three processors shut down during a mission. The algorithms were written in accordance with what we had known to be conservative, and we drew a line that was about 15 percent more conservative above that. It took into account real-time oxygen concentration and then fed that through a head-up display. What that meant in terms of an operational situation is that when a yellow light would come on while you were in the middle of a dive, it would mean that you had to be considering what your decompression scenario was going to be like when you started to come back up. When we got to those stages, we would automatically boost the oxygen concentration, such that we were far in excess of what would be required for a conservative decompression, yet below central nervous system (CNS) toxicity limits. We have never had a bends hit on any of those deep caving projects, even though we're diving at altitude. All of those PLSS units have triplex digital depth sensors that are also sensing atmospheric pressure at the altitude at which we were diving. That was all really taken care of pretty conservatively.

    As far as I understand it, and I don't profess to be an expert at this, the issue of space radiation is really one of shielding and stochastic analysis of what the radiation environments are. You don't want to be out there on the 11-year cycle at solar max. If I remember correctly, there were some analyses of deep space radiation loading that were done in the '70s that indicate—and I can't remember a specific citation, maybe Jim knows—that between two of the Apollo flights there was a solar anomaly of sufficient magnitude that if you had been flying to the Moon during that time, there may have been some serious exposure issues. To me, dealing with deep space radiation exposure is actually something you can mitigate through a series of water barriers and things like that. I believe it's 10 centimeters of water that is sufficient to stop most of that. If you check through the right areas and within NASA, and probably JSC [Johnson Space Center] has people in this, you'll find the answers to those questions. They've been looked at thoroughly. I don't consider that anything more than an engineering problem.

CHRIS MCKAY: Steve Squyres. Introduce yourself.

QUESTION: Steve Squyres, Cornell University. I want to return briefly to an issue that was raised earlier about how small organizations seem to be able to sometimes accomplish more. If you look at what somebody like Ed Viesturs does or what Bill Stone does, or if you look at a group like Burt Rutan and what they are accomplishing there, it's easy to look at an organization like that and say, "Boy, they're lean, they're compact, they're able to get the job done." You then look at larger government agencies and you sense a difficulty there. But I think there is a wonderful counterexample. The wonderful counterexample was what Jim Lovell talked about this morning. It was Apollo.

You look at Apollo, and you look at Gemini, and you look at Mercury, you look at what was accomplished in those days, and it [NASA] was a huge organization. It was in some ways bigger than the NASA of today. Yet, they got things done.

When I look at that, what strikes me is that there was a common thread through those organizations, and that is that they knew exactly what it was they were trying to do. The level-one requirement for Apollo was stated in one sentence by the President. When Ed goes up a mountain, he knows what his level-one requirements are. It is very, very clearly stated. If you have a common goal that is clearly understood by everybody in your organization, I don't think it matters how big the organization is.

CHRIS MCKAY: Good point, Steve. We have five minutes left. We have time for two questions.

EUGENE RODDENBERRY: Hi. My name is Eugene Roddenberry. My father actually created Star Trek and, working in the industry, I have met a lot of people who have been inspired by the show. It's completely different to work in entertainment, but the people who have been inspired have inspired me to look around the world and meet individuals like yourselves. I think you guys have shown that humanity is able to overcome adversities. And the fact that that inspires other scientists and other people in the field to reach for the stars or down to the depths of the ocean—I think it goes beyond that. You guys inspire fans who have disabilities to overcome those disabilities, people who are in relationships, people who have everyday risks that they need to take. I think the fact that they see that humanity can take these huge leaps—these steps beyond—is important It is very impressive. I just wanted to say thank you to everyone. It's very exciting.

CHRIS MCKAY: Thanks for your comments. The next time you send a message to your Dad, thank him for me. I was one of those Trekkies that got inspired to seek out new worlds and all that stuff from watching the show. It's the only television I watched. Larry?

93

LARRY LEMKE: Larry Lemke, NASA Ames. I would like to actually follow up on Steve's comment, and some others on this whole question of how do you take the lesson from small-scale exploration and put it into a bigger context. I agree with Steve. I actually don't think that size of the organization per se is the issue. If you look at the participants on the panel, one of the sort of obvious characteristics that they share is that, not only do they plan the expeditions, they do the expeditions. I am reminded of that comment that there's nothing like the prospect of being hanged in the morning to concentrate your efforts. If you look at the way NASA typically plans a large human exploration project, it is sort of notable that the people who are making the decisions very often do not really have to experience the personal consequences of those decisions. I think if that were to change, then the results might change as well.

CHRIS MCKAY: So, you're saying we should send a NASA administrator to Mars? Is that a way of interpreting what you said?

LARRY LEMKE: Or have the NASA administrator to actually design the vehicle. Get the opportunity to fly in what you design.

CHRIS MCKAY: We actually have a few more minutes.

QUESTION: Bruce McCandless, two Shuttle flights [astronaut]. We get back to the human versus robot trade-off. Currently, when you are looking at a Mars mission, you're looking at speed of light transit times, round trip up to 40 minutes. So, obviously, it's gonna be hours when you get out to the vicinity of Jupiter and Saturn. The thing that seems to be missing is the refined decision-making ability, the ability to adapt to unforeseen situations, to recognize something that you haven't been programmed to recognize. I assume that, eventually, computer science will advance to the point where we can send androids. I wonder if anyone would like to comment on the speed of light transit time as a factor in trading off between humans and robots.

CHRIS MCKAY: That's a good point. Maybe I ought to add to that question, how will that influence the autonomy of an expedition on Mars versus the autonomy of, say, a Shuttle or a station where there is virtually no delay—see them as being more autonomous or not. Does anybody want to approach that? Penny? Dale?

PENNY BOSTON: Yes. I think that it's a return to some of this historical stuff that Jack Stuster was talking about this morning: the fact that those expeditions, before there was the kind of instantaneous communication that we all have, were able to operate and do what they needed to do. We seem to somehow believe now that we cannot do that any more. I think that a different kind of planning within NASA perhaps will be necessary to take us back to some elements of that kind of self-contained expedition.

SESSION TWO

sea

Diving Shipwrecks

I have a fantastic job working for the History Channel. I travel around the world, I get to talk to some very interesting people, I get to talk about history, and I get to dive shipwrecks. This winter, I think I am going to be in France, Scotland, Croatia, the Dominican Republic, and the South Pacific. And when I'm not addressing an audience full of astronauts, I say I have the best job in the world.

Prior to my working for television, I spent more than 20 years working as a commercial diver, largely in and around New York City, where I worked on everything from nuclear reactors to bridges to pipelines—wherever the work was. That was my day job. Before I even got involved in commercial diving, I was diving shipwrecks for recreation. I was attracted by the history, and I was attracted by the challenge that wreck diving afforded me.

There is a big difference between commercial diving and scuba diving. In commercial diving, the diver is a cog in the machine. He is part of a bigger team. When it comes to scuba diving, you are everything. You are your own dive planner and your own dive support. You are your own dive rescue. There is a certain freedom, and, of course, that's linked with responsibility. Eventually, as I acquired more and more experience, I started diving deeper and

John Chatterton
Professional Diver

John Chatterton spent more than twenty years working as a commercial diver and as a boat captain. His passion, however, has been researching and diving shipwrecks. In 1991, his discovery, and then subsequent identification of the German submarine *U-869*, in 230 feet of salt water (fsw) off the coast of New Jersey has been the subject of several television documentaries and now a bestselling book by Robert Kurson, *Shadow Divers*. His diving credits include more than 150 dives to the passenger liner *Andrea Doria* (250 fsw), the first trimix-breathing expedition to the *RMS Lusitania* in Ireland (300 fsw), and the first rebreather dive to the *HMHS Britannic* in Greece (400 fsw). John has worked on numerous projects for television and is currently hosting the television series, *Deep Sea Detectives*, on the History Channel.

more challenging wrecks. I found myself going deep inside wrecks like the *Andrea Doria*. The goal of these dives was just to go where other men had not yet been.

In 1991, Captain Bill Nagle got a set of coordinates from a fisherman 60 miles off the New Jersey coast in what we were told was about 200 feet of water. We put together a trip to the site, we went out there, and what we found was a wreck in 230 feet of water. It was a submarine, later identified as the German U-boat *U-869*. It was a submarine that no one was aware of, where it was, or that it existed. No government, no navy, no historian, no expert could tell us which submarine this was. What an irresistible mystery. It afforded the divers who discovered it the opportunity to rewrite a page of history. We thought at the time that it was going to be a matter of a day or two—on the next dive we would be identifying this submarine. Of course, that didn't happen. It took six years to positively identify it.

In retrospect, looking at our plan, we broke it down into three divisions: economics, operations, and psychology. Economically, we had no financial assistance. We had no support. We had no budget. Essentially, I was going to have to do it on my lunch money. That meant that we were going to dive the wreck to try and identify it the way we had been diving it—as scuba divers. It's a minimalist approach, and it is extremely risky. It's dangerous. Operationally, what was our plan? Well, there were certain legalities that needed to be addressed, dealing with the German government. We then had to do research. Of course, research is what fueled our dive plan. What was there on the wreck site that we could recover that would positively identify the wreck?

The teamwork that we used was indirect. In other words, we would work with one another on research, we would work with one another on planning and coordinating, and that kind of thing. However, you can see that in an environment like this one, to put two or three divers in there is counterproductive to making the dive safer. Because of the silt, because of the very tight spaces in there, and because of the entanglements, you couldn't get in there with more than one person at a time. Specifically, the risks that we were facing relative to the diving were decompression sickness, the possibility of oxygen toxicity, and equipment malfunctions or failures. When we started diving the wreck, we were diving it on air, and we quickly converted to tri-mix with nitrox and oxygen decompression. We had to use redundant systems for primary systems. We also had to be very conscious of health problems. If you faint out in front of this building, they are going to call an ambulance, and they are going to come and get you and take you over to the hospital. If you have a medical problem deep on a wreck, you're going to have a difficult time surviving.

On the wreck itself, it's dark. There are entanglements everywhere. You can see there are hanging wires and that sort of thing. There are fishing nets. There is also the possibility of entrapment, of a loose piece of wreckage collapsing onto the diver. That happened to at least two divers, me being one of them. I'm the only one that survived. You can get lost, either inside the wreck or outside the wreck.

And, then, there's the possibility of panic. The thing that panic does in a very stressful situation is, all of a sudden, your decisions and your actions are not logical. They are not in your own best interest. My goal in this dive was to find a small pad on those hatches at 12 o'clock, which we knew existed. The problem was [that] it was made from white metal that completely [had] corroded away.

Psychologically, this is an extremely intimidating environment. Aside from the fact that 58 German sailors lost their lives inside this submarine, a total of three divers lost their lives diving the wreck while I was working there.

So, you have changing conditions. You are diving by yourself. You also have to consider how obsessed you are, how driven you are. Is this affecting your good judgment? We talked about this yesterday on the panel: when do you abort the mission? You have to be able to do that while you still can.

Six years later, I brought out a tag that positively identified the wreck as U-869. The CBS program NOVA did a two-hour documentary on it. The people that I worked with on that documentary later introduced me to the History Channel where I now work. Robert Kurson saw the documentary and wrote the book *Shadow Divers: The True Adventure of Two Americans Who Risked Everything to Solve One of the Last Mysteries of World War II*. Now Twentieth Century Fox has

> WHY GO THROUGH ALL THIS? MY EX-WIFE USED TO ASK ME THAT ALL THE
> TIME. AND I DIDN'T HAVE A SNAPPY ANSWER LIKE GEORGE MALLORY. IT HAS
> TO DO WITH CHALLENGE. IT HAS TO DO WITH PERSEVERANCE. IT HAS TO DO
> WITH WHO WE ARE, NOT JUST AS INDIVIDUALS, BUT, REALLY, AS A CULTURE.

bought the rights to the book, and Bill Boyles, the man who wrote the screenplay for Apollo 13, is working on the screenplay as we speak.

Why go through all this? My ex-wife used to ask me that all the time. And I didn't have a snappy answer like George Mallory. It has to do with challenge. It has to do with perseverance. It has to do with who we are, not just as individuals, but, really, as a culture. Exploration is very much who we are, and we really have two choices. We either continue on a path of exploration, or we just quit. Not everybody is comfortable with quitting. Certainly explorers aren't. As an added benefit, I am going to close with this letter. I get letters like this occasionally. This one came last Friday.

"My name is Anka Hartung. My grandfather was Mr. Eric Poltey. He was the machinist [obergefreiter] on the submarine *U-boat 869*. As fate might have it, my family and I saw by chance your film about the submarine *U-869*. We are totally moved that we now finally know where our grandfather lies. You and your team have done an awful lot for the families of the lost men. Three people died and you yourself have often risked your life in order to bring certainty and peace into our lives. My grandmother is unfortunately no longer alive to share these feelings with us. You and your team have done so very much for Eric Poltey's relatives, and we sincerely thank you from the bottom of our hearts."

Deep Ocean Exploration

Thanks to all you explorers out there for coming to talk about the wet part of the universe. This conference, of course, is dedicated to the concept of risk. And maybe there's an underlying message about why expose real live human beings to certain obvious dangers when you could—and maybe should—send a machine? Well, I love machines. I mean, I have had a hand in building quite a lot of them, developing and using hundreds of variations on the theme of little machines that operate remotely, as well as those that take a few real, live people inside. And when a job is right, I do believe that it's obvious—you know, pick up a robot, send it, and enjoy it, such as when you're exploring deep under the ice in the Antarctic or in the high Arctic. Send a robot first to check out what's down there, before you go look for yourself up close and personal. I whole-heartedly endorse the concept of using whatever tool does the job, but I think I share with maybe everybody in this room the belief that there's nothing like being there, right? If you can actually get there, why not?

But what about the risk? I'm asked about that quite a lot. You know, why do you do the things that you do? Aren't you scared? Aren't you concerned? I mean, you have a family; don't

Sylvia Earle
Founder and Chair, Deep Ocean Exploration and Research, Inc.

Sylvia Earle is an oceanographer, marine botanist, ecologist, and writer. A pioneering aquanaut and marine explorer, Earle made her first scuba dive at age 17. She has since set the women's depth record for solo diving (1,000 meters/3,281 feet) and logged more than 6,000 diving hours—feats that garnered her the moniker "Her Deepness." The author of five books and numerous scientific and popular articles, Earle tirelessly calls for the preservation and exploration of the world's marine ecosystems.

they object to the idea of you going down underwater? It's dangerous! My answer is usually the same. The most dangerous thing I do almost every day of my life is to get into an automobile, get on the highway, and move along at reasonably moderate speed, and I face traffic coming the other direction, and the only thing that keeps me from banging into that traffic is a painted line down the middle of the road and a mutual desire—I hope it's mutual—to live. That's really dangerous.

I think about explorers of the past and what they would think of this conference. I mean, we are so obsessed with safety these days, so obsessed with risk. Can you imagine what OSHA would say about Christopher Columbus, or about the *Challenger* [oceanographic] expedition in 1872 as they made their preparations to go for four years around the world, going places where nobody had been, exploring deep parts of the ocean? Imagine what they would say about William Beebe with his little bathysphere and Otis Barton, the engineer [who created the bathysphere]. And if you've seen any of the films—and I have—of their operations, anybody associated with OSHA would have heart attacks just watching. No hardhats! No hard shoes, running around barefoot on the deck with this heavy equipment being slung around. Who would have insured Beebe or his machine back in the 1930s? There's something that's happening to us as a species as we become risk-averse.

But I share with Anne Morrow Lindbergh some thoughts about risk. She and her husband Charles paved the way for the first flights across the North Pole, looking for ways to establish new commercial air flight routes back in the 1930s. And when asked by a reporter as they set off for their first flight across the North Pole—north to the Orient—the reporter asked her, "Can't you even say that you think it's an especially dangerous trip?" And she said, "I'm sorry, I really don't have anything to say. After all, we want to go. What more is there to say?" And that's it. You know, as explorers, like little kids, we want to know what's around the next corner, what's under the next rock, what's over the next horizon, what's in the deep, what's beyond the next star—or starfish.

Danger is the silent partner of exploration, no doubt about it. But just try to avoid risk in everything you do. I have a home in Florida—that's risky! I have a home here in California—think of the earthquakes—that's pretty risky. I live in this day and age. I walk in the streets of Washington, DC at night! That's really risky.

When it comes to the ocean, I want to go. I want to have access, not just to the highest reaches of this planet. In fact, since the first ascent to the top of Mount Everest half a century ago, more than 2,000 people have been to the top of Mount Everest—literally the top of the world. It will soon be half a century since the first successful trip to the deepest part of the ocean. That was the Everest of the ocean, 11 kilometers down—7 miles—the bottom of the Marianas Trench, not too far from the coast of the Philippines. That was nine years before the first footprints were on the Moon—1960 when that took place—13 years after Thor Heyerdahl's expedition across the Pacific with a balsa wood raft. Again, OSHA would not have approved. At a depth of seven miles, two men looked out of the port of the little machine, the bathyscaphe *Trieste*, at a depth of seven miles and a pressure of 16,000 pounds per square inch, in that eternal darkness of the deep sea, except for bioluminescent creatures, which are virtually everywhere in the ocean. They saw eyes looking back. It was a flounder-like fish. And everybody joked, of course, it had to be a flounder-like fish, a flat fish, with 16,000 pounds of pressure per square inch.

But there you are. For about half an hour, almost half a century ago, they had a glimpse of the deepest part of the ocean. Nobody's been back since. How can this be? Presently there are four vehicles that exist that can take people to just over half the ocean's depth—the two Russian Mir subs, the French *Nautile*, the Japanese *Shinkai 6500*. The Japanese tethered robot Kaiko did get some observations a few times in the deepest part of the sea in the last decade, but it was lost at sea last year. They confirmed, however, the existence of abundant and

YOU KNOW, AS EXPLORERS, LIKE LITTLE KIDS, WE WANT TO KNOW WHAT'S AROUND THE NEXT CORNER, WHAT'S UNDER THE NEXT ROCK, WHAT'S OVER THE NEXT HORIZON, WHAT'S IN THE DEEP, WHAT'S BEYOND THE NEXT STAR—OR STARFISH. DANGER IS THE SILENT PARTNER OF EXPLORATION . . .

diverse life at the deepest part of the sea, and soon, Woods Hole Oceanographic Institution will have, with support from our taxpayer funds, a tethered robot that will, again, go to the deepest sea. But it will take a few years. China is building a 7,000-meter manned sub, and the United States is getting back into the deep sub game when that workhorse of all subs, the *Alvin*, will be replaced in the next few years with a 6,500-meter sub.

Well, I say, why only 6,500, why 7,000 meters when we're looking at an ocean that is 11,000 meters deep? I want to go to the deepest part of the ocean. I mean, who doesn't? Why wouldn't you want to go? But I'm told, you know, we've got access with a 6,500-meter or even a 7,000-meter sub to about 98 percent of the ocean. So, it's only 2 percent, why worry about that? Well, it's 2 percent—it's an area about the size of the United States and an area about the size of Australia or China, and we'll just write that off. And it's a unique high-pressure realm. Remember, 16,000 pounds per square inch of pressure. Where else on the planet are you going to find forms of life that can survive in a realm like that? It's a place where basic ocean processes are taking place as well, the bottom of the deep trenches where the crust of the ocean is diving under the continental plates.

Well, I've conveyed my concerns about the powers that be that are stopping at 6,500 to 7,000 meters. I say, "Lewis and Clark didn't stop at the Rockies and say, 'That's good enough. Why bother going all the way to the coast?' Sir Edmund

Hillary and Norgay Tenzing didn't stop 98 percent of the way up to the top of Mount Everest, and we didn't travel 98 percent of the way to the Moon and turn around and say, "That's good enough." Or to Mars. You know, we actually have sent probes and landed on Mars, and someday we will get back to the deepest part of the sea. And, frankly, I don't know what's stopping us. Unless there's a certain resistance called risk.

I, like Anne Lindbergh, like many of you here, I suppose, really do want to go. And here's the thing. I'm far more concerned about not taking the risks involved with exploration than risks that are involved with doing what we are doing. I mean, suppose we just get ultrasafe and stay in bed—that's risky too. As an ocean scientist, as chief scientist of NOAA back in the early '90s, they started calling me the "Sturgeon General" because I expressed concern about what was happening to the planet. This is, after all, our life support system. And, as any astronaut will tell you, you learn everything you can about your life support system, and then you do everything you can to take care of your life support system. And we haven't learned a great deal yet about our own life support system. This blue planet—less than 5 percent of the ocean has been seen, let alone explored. And I don't think the risks are really worth talking about when you consider the gains and the risks of not taking whatever modest risks there are out there.

I am concerned about the health of this planet—our life support system—starting with the Earth's blue heart, the ocean. I think of the ocean as the engine that drives climate and weather, regulates temperature, generates most of the oxygen, and absorbs much of the carbon dioxide. It's home for 97 percent of life on Earth, and that's not surprising considering that that's where 97 percent of the water on Earth is. As Chris McKay—one of my great heroes—says, "Water is the single non-negotiable thing that life requires." Huh! There it is.

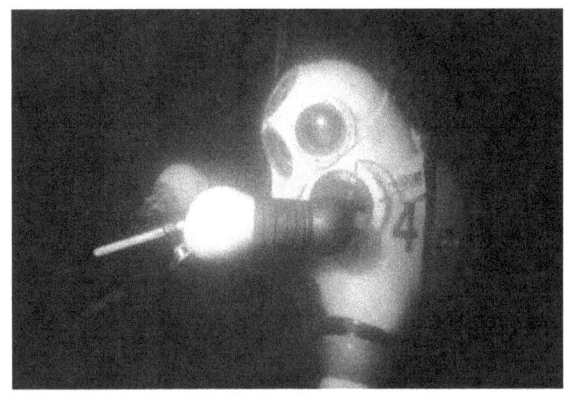

In the past half century, we've learned more about the ocean than during all preceding human history, but it's not good enough—there's so much more that we need to know. And, at the same time that we've learned more, we've lost more. In the last half century—the last half century!—90 percent of the big fish in the ocean have been extracted. Ninety percent! Think of it. Half the coral reefs are either gone or they're in really a sharp state of decline. Kelp forests from Tasmania to Alaska are not in the same good health that they were 50 years ago. They, too, are in a state of decline. I hope you enjoyed that tasty bit of halibut that you had last night—those of you who consumed it—because they're among the big fish—along with tuna, sharks, swordfish, grouper, snapper, California rockfish—that have plummeted in my lifetime, in your lifetime, because we are so good at extracting things from our life support system before we even understand how it works.

So think about what the risks will be of not taking the relatively small risks involved in exploration today. The chemistry of the planet is changing. What does

THIS BLUE PLANET—LESS THAN 5 PERCENT OF THE OCEAN HAS BEEN SEEN, LET ALONE EXPLORED. AND I DON'T THINK THE RISKS ARE REALLY WORTH TALKING ABOUT WHEN YOU CONSIDER THE GAINS AND THE RISKS OF NOT TAKING WHATEVER MODEST RISKS THERE ARE OUT THERE.

that mean to the little critters that are out there? Especially the microbes that really dominate the way this planet works? Won't take much to set off a whole new suite of events based on the changes in chemistry that are taking place now. Our security as a species is at risk for our reluctance at not taking the relatively small risks involved with what some regard as cutting-edge exploration. As never before, we really do have a chance to get out there and make a difference—and maybe as never again. I want to show you now something to cause you to dream with me about what the potential is. Why aren't we out there in the ocean? Why aren't there fleets of little submarines like there are fleets of aircraft up in the sky? There is a little one-person sub called "Deep Worker", built up in Canada. There are, I think, about fifteen or sixteen of them in operation now around the world. For five years as the explorer-in-residence—what a cool title!—at the National Geographic Society, I had the chance to engage more than a hundred people—scientists, teachers, administrators, paper-pushers, economists—to learn how to drive those little subs. They're so simple to learn how to drive that even a scientist can do it.

And we did it, looking at the coastline of the United States, focusing on the small but promising counterpart to the national parks on land—marine sanctuaries. There are a few. It amounts to less than one percent of our coastal waters, but, nonetheless, we've made a start toward protecting our life support system around this country. By getting into one of these little subs—one atmosphere, no decompression—we could go as much as two thousand feet. It's a start toward the ultimate 35,800 feet—the deepest part of the ocean, 7 miles. Why shouldn't we invest in fleets of little subs that can take anybody who wants to go for whatever reason? Whether you want to write poetry or whether you want to write a business plan or whether you're an explorer interested in science, this is the major part of our planet. It's blue! It's water. And it's largely still inaccessible.

I was among the first in this country, back in the early 1950s, to enjoy using one of the first aqualungs that first came into the country. I salute Jacques Cousteau almost every day for giving me a passport into the ocean, and I love the concept of being able to fly freely in the sea as a diver. And that's what these little subs do, too. As a diver, all by yourself, people say, "Aren't you afraid all by yourself?" Well, again, what else do we do all by ourselves? I love subs of all sorts: 1 person, 2 person, 6 person, 30 person, or passenger subs that take people out into the sea at least down to 50 meters or so these days. What is stopping us from gaining access to anywhere in the ocean we want to go? Anytime we want to go? We need to understand what's out there, what's down there. This is a moment in time—a crossroads in time—when we know that our life support system is in trouble. This part of the solar system is changing, this blue planet, this Earth. With all due respect to our goal of going elsewhere in the solar system to set up housekeeping—and I love the idea of going to Mars, I'd love to be able to go myself and come back—the fact is that, look as far as we might, the Earth is the place that, for the foreseeable future, we have got to come to grips with and take care of it. That's really what is at risk: our future.

Ocean Futures

Ladies and gentlemen, it is a great privilege and honor to be here, and very humbling, knowing who is here in this audience. And many of you I've had the opportunity to meet, and I have a lot of respect for what you do.

The symposium's invitation states that NASA was created to pioneer the future. I will always remember the difficult times of trying to sell television programs with some of the networks in the United States. The people who were putting up millions of dollars were asking my father, "So, Captain, what do you expect to find?" And his answer to those people who were about to make major commitments was, "If I knew, I wouldn't go."

This extraordinary desire to see what's on the other side of the hill is what has animated all of us. This cannot be done if we do not have a commitment to preserve and protect the resources of the present. It is a dream as old as consciousness to explore the stars, so we must continue to explore, but with an equal commitment to protect the quality of life on Earth, which we are not doing. It will do no good to send people into space or underwater if it becomes an escape from intolerable conditions here at home. That being said, as famous a pioneer underwater as my father was, and his team, they took risks they didn't even know

Jean-Michel Cousteau
President, Ocean Futures Society

As an explorer, environmentalist, educator, and film producer for more than four decades, Jean-Michel Cousteau has used his vast experiences to communicate to people of all nations and generations his love and concern for our water planet. The son of ocean explorer Jacques Cousteau, Jean-Michel spent much of his life with his family exploring the world's oceans aboard *Calypso* and *Alcyone*. After his parents' deaths in the 1990s, Jean-Michel founded Ocean Futures Society in 1999 to carry on this pioneering work. Responding to his father's call to "carry forward the flame of his faith," Jean-Michel's Ocean Futures Society, a nonprofit marine conservation and education organization, serves as a "voice for the ocean" by fostering a conservation ethic, conducting research, and developing marine education programs. Jean-Michel has produced over 70 films and been awarded the Emmy, the Peabody Award, the 7 d'Or—the French equivalent of the Emmy, and the Cable Ace Award.

existed, compelled by the adventure of what they were seeing for the first time. They took those risks because they were inspired by the importance of the realm they had entered, just as space explorers were, are now, and will always continue to be. Having seen the world underwater, my father then dedicated his life to protecting it. He also came to appreciate that everything is connected, and, thus, he became concerned about the water systems of the planet, the land, the atmosphere, and the quality of life for people.

THE PEOPLE WHO WERE PUTTING UP MILLIONS OF DOLLARS WERE ASKING MY FATHER,

"SO, CAPTAIN, WHAT DO YOU EXPECT TO FIND?" AND HIS ANSWER TO THOSE PEOPLE WHO

WERE ABOUT TO MAKE MAJOR COMMITMENTS WAS, "IF I KNEW, I WOULDN'T GO."

I think NASA is in the same position relative to its view of life on Earth from space. When my father pushed me overboard at the age of seven, I had a tank on my back and, in those days, children did not argue with their parents, so I've been a scuba diver ever since. Some of my earliest views were formed in the middle of the night, when my father would wake me and my brother out of a sound sleep to stand on the terrace in our south-of-France home to look at the sky, full of stars, planets, and the Moon. We were learning about nature firsthand. Jacques Cousteau was a dreamer, full of excitement to explore outer space when it was only the subject of science fiction at the time. Fifty-nine years ago, he pushed me overboard.

I think my father and his team were willing to take great risks, risks they realized they couldn't even describe or predict, because first, looking up at the stars and then into the oceanic abyss, they knew the greater risk was ignorance. This is as true today. Our invitation also asks, "Why are sacrifices made in the name of exploration more notable than the losses incurred in the course of everyday life?" I think it is the nature of our species to focus on drama. We don't accept short-term, immediate, dramatic risks, but long-term, slow, less dramatic yet more important risks we ignore—i.e., species lost, pollution, and reducing the habitability of the planet for life. We get excited about lives lost from short-term, dramatic events, but are oblivious to thousands of people losing lives from the demise of the environmental system that provides them with income, food, and a quality of life.

For example, in the U.S., it is estimated that the amount of oil runoff flowing from urban pavements into the oceans creates the equivalent of an Exxon *Valdez* every eight months, as reported by the Pew Ocean Commission. Yet, not a word reaches the masses, and even if it did, there would be little outcry. Even the fact that six thousand children die every day from lack of access to clear water creates no outrage. We seem to accept, even ignore, these pernicious risks. I think it is the duty of those of us privileged with the ability to explore to point out both the dramatic and the mundane, but certainly more significant, events.

So, how can we look with vision and commitment into the future of space travel? I think we have to do it by mounting rearview mirrors on our spacecraft. By that, I mean that, while moving farther into space, we simultaneously take the opportunity to include equipment that will continue to monitor with greater sophistication the state of the Earth. Basically, we cannot fulfill our dream of exploration in outer space or inner space if home base is unlivable. NASA is powerfully positioned to create what I call the Global Ocean Network, which at our Ocean Futures Society we have started working on in a conceptual phase, whereby it would be a way to constantly monitor from space with an array of vessel buoys, habited buoys, drifting buoys, whatnot, both bringing the dramatic events and long-term trends in the planet's water system. As Sylvia just said, it is our life support system.

This is nothing new. I have a report right here, given to me a few days ago, from a 1971 meeting of my father at NASA Headquarters with Dr. Wernher von Braun and NASA officials. My father presented the case for a global monitoring system "to monitor the primary production of life in the ocean and to monitor the deterioration of life in the ocean resulting from human activities and from natural forces..." His dream was for NASA to launch satellites to monitor sophisticated ocean sensors. Much has been done in this direction, but now it needs to be part of every endeavor. We need to take an aggressive marketing and public relations approach to selling the future and the risk to the public, something we've not done well. We need to engage them in realistically assessing risk and prioritizing issues. We need to motivate and mobilize them to take personal action and political action to ensure we have an acceptable future for our children. The future based on the direction we are [currently] headed is unacceptable.

NASA is in an unprecedented position to participate in necessary new directions. Infusing future space exploration with stewardship of our planet, we will accomplish two things that have to do with risk: We will have upped the ante in terms of what we can gain by risking human life to further our knowledge, and we will have shown our regard for that human life by protecting it in the only place we know it to exist. There will always be brave men and women willing to risk their lives for exploration in outer space and underwater. We need to dignify their courage, and possibly deaths, by making sure we are doing everything to protect not only their lives, but the life-giving system of the planet through their work as well.

I'd like to tell you why we take risks. This incredible planet of ours, the only one with sufficient quantities of water that we know of to have the kind of sophisticated life like we have, has inspired a lot of people like my dad to pioneer.

They didn't know what they really were doing. They were very cautious. And most of them, anyway, stayed alive. But it was touch and feel. It was this unbelievable curiosity that animated them and opened the ocean world to millions of people.

In South Africa, I was taught to dive and hang on to the back of the dorsal fin of a 14-foot great white shark. Was I taking risks? Very calculated risks, much less than when I cross Fifth Avenue in New York. But in the process, we're making people understand that these animals are part of our system.

Was I taking risks when I wanted for the first time to go down with a ship, sink with a ship, a Russian frigate that was made into a dive site? I always wondered, what happened in the minds of those people as they sank with their ships, the captains, the people in charge? A few months ago we were in the middle of the Pacific working on an island, Laysan, where nobody lives, and finding all our refuse. Fifty-two countries were represented there with probably tens of thousands of tons [of refuse] just lying there with fishing nets and debris. We are using our ocean, we're using our own home, as a garbage can, a universal sewer. At some point nature will say, I can't handle it anymore, and we are getting signs of that today.

I believe that exploration and taking risks is what is going to change the face of the planet today. We have new equipment, free breathers, new fins designed by imitating the flip of an Orca, new lights, new submersibles, new communication systems which, as Sylvia just rightfully said, will allow us to explore not just the five percent we've explored, but a hundred percent. And that's what's going to make us do the right thing. Because how can we protect what we don't understand?

So this risk we're taking is for the bettering of the quality of life for the human species on the planet. Those sharks we were diving with at 200 feet of depth, they don't care. We do. We want them to stay there just like anything else.

I will never forget the comparison that my dad made one day when he told me, "You know, the planet is like an airplane with wings. Every time you remove a rivet you are removing a species. At some point, it may just collapse." We don't want to go there.

And the decisions that our brains, that our industries, and political representatives anywhere in the world will make will allow us to fulfill our dream and take calculated risks. And that, I believe, is what animates every one of us here. I have no job, I have a passion, and I will not retire until I'm switched off.

Exploration and the Risk-Reward Equation

Defining and controlling risk in exploration operations is a tough and continuous challenge that requires the application of a range of methods from the qualitative to the quantitative, and, ultimately, to be successful, requires consistent application of informed good judgment.

We've already heard a lot of very insightful themes from the previous speakers, and you're going to hear some of them again from me, but possibly from a slightly different perspective. For my entire adult life I've been in relatively risky professions, starting out as a scuba diving instructor and boat captain in the Caribbean, where your job is basically to keep people from killing themselves, and you see it all, from people who sit on sea urchins and scream and spit out their regulator and their false teeth, and then go shooting to the surface, to people who go chasing after aggressive tiger sharks. So you learn to expect that anything can happen.

After that, I worked as a commercial deep-sea diver doing subsea construction in the offshore oilfield, and then, later, as vice president of the world's largest subsea contractor. Currently, I'm an astronaut, and involved in doing high-risk human research for space decompression procedures.

Michael L. Gernhardt, Ph.D.
NASA Astronaut

NASA selected Michael Gernhardt as an astronaut in March 1992. His technical assignments have included development of nitrox diving to support training for the Hubble Space Telescope repair and a variety of Space Station extravehicular activity (EVA) developments; spacecraft communicator (CAPCOM) at Mission Control Center, Houston, during various Shuttle missions; and leading an international research team in developing a new exercise prebreathe protocol that improved the safety and efficiency of space walks from the International Space Station. Gernhardt presently serves as a member of the astronaut office EVA branch and as principle investigator of the Prebreath Reduction Program and manager of Johnson Space Center's Environmental Physiology Laboratory. A four flight veteran, Gernhardt has logged over 43 days in space, including 4 spacewalks totaling 23 hours and 16 minutes. He was a mission specialist on STS-69 in 1995, STS-83 in 1997, STS-94 in 1997, and STS-104 in 2001. Gernhardt is assigned to the crew of STS-119.

As a professional, it's important that you address the risk-reward equation. Basically, that equation states that the utility, or the degree of success, is equal to the probability of success times the reward, minus the probability of failure times the cost. As a professional, if you don't balance this equation properly and end up most of the time with a really large, positive number, you're either not going to live very long, or, if you're in business and you're killing your people, you're not going to stay in business very long.

Commercial diving is potentially a very dangerous business, but, in fact, it's actually safer than many forms of nonprofessional scuba diving, because we understand that it's risky and we plan for those risks. Some of the nonprofessionals tend to focus more on the reward component of this equation. It would be really neat to dive on this wreck or really great to go in this cave. And, unfortunately, they don't understand the risk side until it's too late.

As individuals involved in these operations, it's vitally important that you understand and accept the risks that you're getting into. And it's also important that the individuals have direct control of the risks through their own actions.

Commercial deep-sea diving is potentially very dangerous. Some of the work that we do includes very complicated construction tasks that would be dangerous on dry land. An example is a hyperbaric welding job, where, in order to do code-quality structural repairs of offshore platforms, we actually have to weld in a dry environment, because in wet welding, the water quenches the weld so fast you get hydrogen embrittlement. So we have to design these multipiece habitats that we have to install around the tubular truss structure of the platform, install seals, dewater the habitat, and then go inside and weld in a dry environment.

These are challenging operations at very high forces. A lot of time you're working in current conditions, at close to maximum aerobic capacity. These operations would be dangerous on dry land, but we do them at depths of up to 1,000 feet, under extreme physiological stresses, working in a dynamic, harsh environment that is capable of radical changes over short time periods. And many times, you're working in limited or zero visibility on the muddy bottom. And, so, you've got to realize that that's risky, and plan and address those risks.

In my mind, I divide risk into two categories. There's what I would call the corporate or programmatically controlled risk, and these risks relate primarily to the design of the equipment, the degrees of redundancy, the reliability, things of that nature. An example is a saturation, helium-oxygen saturation, diving system. And if you're not familiar with saturation diving, we use this method to increase the efficiency of the amount of bottom time we get for the amount of decompression time. If we were to work at 500 feet for 30 minutes, it would take over 24 hours to decompress. Once you stay on the bottom 24 hours or longer, the partial pressure of inert gas in your inspired breathing mixture comes to equilibrium with the tension of gas dissolved in your blood and tissues. Then your blood and tissues will not uptake any more inert gas, and it will take 5 days to decompress, whether you stay on the bottom for another minute or another month.

So, we actually live in these pressurized habitats on the deck of the vessel, breathing a helium-oxygen mixture. At 1,000 feet it's less than one percent oxygen. Then we transfer under pressure into a diving bell, we make a seal on the bell, the bell is deployed overboard, and acts as an elevator to transport the diver to the subsea worksite. Then we lock out of the bell, do eight hours work, reenter the bell, and make a seal-return to the surface under pressure where we transfer into the living chamber on the deck of the vessel or platform. With a six-person crew, we do 24-hour-a-day operations. That results in about 20 hours of working bottom time per 24 hour-day. We spend a month under pressure, so the working-time to decompression-time ratio is about five, compared to the surface diving position where you're less than point one.

This is a very efficient form of diving; it's very challenging with respect to the life support systems. Minor changes in the oxygen percentage can mean the difference between hypoxia and acute oxygen toxicity. Same kinds of limits on the carbon dioxide. Temperature and humidity are very sensitive at these extreme pressures and gas densities. A temperature swing of a few degrees, a few percentage points of relative humidity change, is the difference between comfort and discomfort. And larger swings than that are life and death.

On top of that, we're locking out, we're working in an oilfield environment, where you can bring trace contaminants back into the habitat. So all this has to be accounted for ahead of time, and controlled, and if you do a good job at the corporate level, the equipment and procedures are safe, and you're happy to go use it. On one of our diving support vessels, we actually have a 16-person saturation habitat built in below decks. It's very much like a space station, with living quarters and node. There's a thing called a moon pool, we deploy the bell through the bottom of the vessel. So we do all kinds of very challenging operations, including some very unique decompression procedures.

I had the opportunity to work with a man named C. J. Lambertsen, who actually invented the oxygen rebreather, and is considered the founder of the Underwater Demolition Team (UDT). He actually worked for the OSS in World War II, and he was the medical director of the company I worked for, and I worked closely with him for almost 25 years now. Very wise guy, very smart, very intellectual, very good operator. And his attitude was always, what do we have to do? Now, how do we do it safely? And that's the right question to ask.

The other question is, what can we do safely? And if you ask that question, you don't have the focus. I mean, there are a lot of things you can do safely. You can watch television, you can go bowling. Oops, no, you might hurt your back. So you see where that's going. So it's important to define what you want to do. Then you have clear focus, and you can address the risk and do it safely. If you don't have a clear vision of what you are attempting to do, then its difficult to analyze and control the risks, and, ultimately, you can end up being less safe, even though you start with a more conservative attitude.

The other form of risk that I categorize is what I call the individual or team-controlled risk. And even though the company might provide you with

113

safe diving equipment and methods, we're doing heavy-duty construction in a dynamic environment, and there are all sorts of risks that are directly in the control of the diver on the end of the hose and the topside team supporting him. One example is a platform repair we did off Peru. The platform was falling down, and we had to burn off the old parts of the platform, and then install these clamps and braces and, basically, rebuild the entire platform underwater. A lot of times, we're working in two or three knot currents with heavy surge conditions. In order to rig the repair braces and install them underwater, you've got, sometimes, two crane lines and four air tugger lines that you have to direct in order to transfer the multiton brace into position at the underwater worksite. You're down there in these heavy currents and surge with limited visibility and your life support umbilical [is] potentially in the middle of all of these crane and air tugger lines, and if don't have good situational awareness, you can get your umbilical hose, hand, or arm in the middle of the trajectory of these swinging flanges and lose your arm, or your fingers, or cut your umbilical hose. And, so, you have direct control over these risks.

SO IT'S IMPORTANT TO KEEP FOCUSED ON WHAT YOU'RE DOING,
AND BE CONFIDENT AND AGGRESSIVE, AND UNDERSTAND THE
RISKS AS BEST YOU CAN AND THEN GO DO IT.

I don't know how NASA's safety would quantify this kind of operation. We typically approve things by testing or analysis. I don't know what you test here, because every circumstance is different, impossible to quantify, and, generally, unrepeatable. And, so, these risks are very much in your control, and your skill, and the supervisor's assessment of your capabilities, is the only level of control of these risks, along with very good planning and teamwork.

One of the observations that I've made is that to do this kind of stuff safely, you have to have the right attitude, you have to plan it, and you have to work with your team, your topside team, and the people controlling all these crane lines, and you have to go in with a good plan, and you have to be confident and aggressive.

And the people that I saw getting hurt were the people who had checked out a little bit. Their heart really wasn't into it, and they wouldn't attack the pre-dive planning, and then they'd get in and they'd hesitate at the wrong moment, or something like that, and they would have the accident.

And I think the same observation would be true for an organization. If you become so risk-averse that you indiscriminately apply your resources to controlling trivial risk, then you don't have those resources to apply to the important risks, and you lose your focus, and you really don't accomplish that much, and, frankly, you're probably not that much safer. So it's important to keep focused on what you're doing, and be confident and aggressive, and understand the risks as best you can and then go do it.

We heard a lot of interesting and eloquent quotes from explorers, and I have to tell you one of my favorite quotes was from my first Shuttle commander, a guy named Dave Walker. Dave is no longer with us, but he was a remarkable human being and a great team builder. He actually christened our crew as the "Dog Crew," and he gave everybody a dog name. The only condition was you couldn't like your dog name. So, being a rookie and a diver, I was Underdog. His call sign was Red Dog. Dave said to me, "You know, Underdog, it's a fine line between bleep and bleep hot." I can't say it exactly the way he said it. The bleep starts with an "S." Dave had probably known both sides of that line, so he really understood that. It is a fine line. It is a fine line between being a cowboy and taking too many risks, and then, on the other side, being so risk averse that you don't get anything done and you're not as safe as you should be.

Now, as far as the risk-reward equation and the commercial diving industry, we have to be safe or we don't have a business. It is the right thing to do. You don't calculate that we're willing to lose this many people or anything like that. You do the very best you can to make things safe. You also make them cost effective and efficient. What we have done over the years is, we started out with the divers in a hands-on environment. We have slowly evolved the human back from the direct operational environment. Instead of divers having to go into saturation and incur all these physiological stresses, we had one-atmosphere dive suits. That was one step. We then stepped further back from that with the introduction of remote operated vehicles [ROVs]. I was in commercial diving in the late '70s and early '80s when these became widely used. It was pretty comical at first, because they were way oversold. The salesman would promise the oil companies that you could do all kinds of things. We actually ended up making a lot of money as divers rescuing these things when they failed or got fouled up on a structure. One of the key things that we learned is that it's not so much the capabilities of the human or the robot; it's both sides of the interface, which includes how you design the tasks to be compatible with the diver or the robot. The integration of both sides of this equation results in a work system versus just a diving suit or a robot. What we did was work with the oil companies to reengineer the subsea equipment so that we could work on it easily with ROVs. We ended up actually being able to produce as efficient work with these ROVs today as we could with divers in previous years. An example is what we call the bucket. We actually made the task so simple that the only task was to dock the ROV into this conical interface. We had different tooling packages inside that would do different things, ranging from small and large valve actuations to mating electrical and hydraulic connectors, but to the operator, the task was always the same, dock the ROV into the bucket. So you try to keep it simple.

When you keep things simple, it actually gives you more brain cells to apply to situation awareness to know how to stay out of trouble. We have actually evolved some of these concepts up to the Space Station, with the microconical interface. So, the message here is to keep the task and operation as simple as you can and, ultimately, that makes it safer, because you have more reserve capabilities and situational awareness to deal with the unexpected. When you plan an operation right at the limits of your capabilities, your safety margins go down.

People always ask me, "Was it more dangerous in commercial diving or the astronaut business?" I think the answer to that is that in the space business, getting to the work site is a lot more dangerous than riding the boat out and coming back. But in commercial diving, once you're at the work site and dealing with all these dynamic forces and physiologic stresses, it is probably riskier than doing a spacewalk.

One of the things I think will happen, though, as we evolve to planetary exploration is that instead of training for a whole year to do a spacewalk and having a whole ground team behind you, we're going to be doing EVA [Extravehicular Activity] every day, with a plan that has been developed, at best, the day before and one that is likely to change many times during the course of the EVA. The balance of risk is going to shift between the corporately controlled risk on the redundancy of the vehicle to the personally controlled risk when you're doing these EVA operations. We need to have people who can make good judgments and good decisions in a relatively unstructured and dynamic operational environment.

I participated in one of the first NEEMO (NASA Extreme Environment Mission Operations), which is a program we have going with NOAA [National Oceanic and Atmospheric Administration]. The underwater habitat they have off the Keys is a great analogue. We actually lock out, do coral reef science, and spend nine hours a day in the water. They have remote way stations where you refill your tanks. I have proposed that this is a great analogue to use to parameterize that operational space. How far away from the habitat are you? What are your consumables? How long will it take you to get back to the habitat, and how much air will you consume? You have to make all these real time decisions about when to refill tanks, when to start and stop working, when to head back to the habitat, et cetera. The NOAA team has really tight flight rules. If you come back to the habitat with less than 500 psi or one second beyond your flight plan, you're busted. You're not going to dive anymore. It really builds good decision-making. They have done over 27,000 excursion dives with a perfect safety record.

The notion would be to parameterize this operational space, and then ask yourself the question: If we're going to work on the Moon and we want to explore a 200 kilometer radius, then what life support do we need? How fast do our transport vehicles need to go? Where should the way stations be? There is a lot that we can learn from land and subsea analogues that we should be applying to our mission design well before we set foot on the Moon or Mars.

I am going to transition quickly from subsea to space on the topic of decompression. I will also talk about the difference between qualitative and quantitative risk control. We have to decompress in space because we work in low-pressure space suits, and we want the pressure to be as low as possible so that we have minimal forces and torques across the suits so we can work. We actually have to get rid of nitrogen much the same way a diver does.

On the trials of the Shuttle decompression procedures that we have all used, we had 25 percent decompression sickness. You ask yourself, is that acceptable or not? It turns out there are some things about altitude DCS (decompression sickness) that are much different than diving. If you talk to the divers here, they will say you've got to have way less than five percent. In commercial diving, we had about 0.01 percent. Altitude decompression sickness is different, primarily because you pre-breathe the oxygen and undersaturate your brain and spinal cord, so we don't [have] nearly as many serious symptoms of DCS that we see in diving.

When they did the Shuttle ground trials, they came up with 25 percent DCS, and they had a committee come in and they said, "Well, what do you think? Is this safe or not?" You can find anybody to say it's safe or it's unsafe. It turns out that we have not had any decompression sickness in flight, probably because the ground model was not that accurate. I don't have time to go into all those details, but the point is that it was the assessment of acceptable risk was very subjective.

When I started the pre-breathing production development for the procedure we are now using on the Space Station, I took a whole year with a large team to define what acceptable risk was. I pulled in the Navy and the Air Force, the flight directors, who are great guys, who are really great at analyzing data and making decisions, the flight surgeons, and the astronaut office. When we had the first meeting, I said, "Everybody in this room has an opinion about what acceptable DCS risk is. Recognize it is only your opinion." We proceeded over the course of a year to pull in all the data we could, analyze the data, and when we extracted the last little bit of information out of that, we finally made the decisions.

There was a lot of talk yesterday about staging things. We actually staged into this. You couldn't get a consensus right off the bat as to what acceptable DCS risk was, but I took the tack of saying, what's the highest risk we could have and still build the Space Station? We had a policy that if you had Type I DCS on an EVA and it resolved, you could go EVA again in 72 hours. This was consistent with Navy and Air Force procedures. If you have the second Type I hit on this same mission, then you were out. If you had Type II, serious DCS, you were also out of the rotation.

We then did a Monte Carlo simulation of the entire Space Station assembly and maintenance model, applying this policy and subjecting it to the constraint that we be 95 percent confident that we would always have two crew members to do an EVA. That defined the uppermost risk we could have. We then looked at other factors and actually ended up picking a level of DCS risk of 15 percent at upper 95 percent confidence level, which, [as it] turns out, is below a threshold where there has ever been a report of Type II DCS in our database. We do these trials with human subjects. Subject safety is our number one priority. We have defined very explicitly what the accept conditions are. Even though the research is difficult, it is pretty easy to make a decision, because we have prospectively defined the acceptable risk criteria. You design the experiment, you do the trial, and, if it meets it, you're great. If it doesn't, you reject it and test the next protocol.

Some of these quantitative risk definitions and control techniques would be applicable to other aspects of vehicle and mission safety design. Statistics are a good tool, to be used in conjunction with informed good judgment, not a replacement for it. It's a fine line that we will have to walk as we move forward with the next generation of exploration missions. We will need to understand and accept that they are risky, define clearly what we want to do, define and control the risks as well as we possibly can, and then go do the mission recognizing that we have done everything practical to control the risks, but that we will never totally eliminate them.

Titanic and Other Reflections

I am also honored to be part of this august panel, which includes two of my heroes from the undersea world, and some of the people I'm just meeting today. We live in an age when the land area of our planet has been explored, mapped, imaged, settled, and exploited for whatever it has to offer. It's definitional that what remains to be explored are the most remote, inaccessible, and inhospitable parts of our world, or places that are not a part of our world at all. This basically means that the easy stuff has been done, if you want to consider polar exploration and all the great pioneering work in the ocean the easy stuff. The hard stuff is in front of us, and it means we are now confronting even more hostile and extreme conditions and requiring more sophisticated technology and support systems in order to do our exploration. Correspondingly, we are facing more complex and subtle forms of risk than ever before.

I have lived with risk for my entire professional career as an action film director. I regularly asked people, with a completely straight face, to set themselves on fire, to flip their car over, to leap out of an exploding building, to ride on top of a tractor-trailer truck that's on its side skidding, to fly a helicopter underneath an overpass with two feet of clearance on either side of the rotor tip, and even to ride a sinking ship down underwater.

James Cameron
Writer/Director, Undersea Explorer

Born in Kapuskasing, Ontario, Canada, James Cameron grew up near Niagra Falls. In 1971, he moved to Brea, California, where he studied physics at Fullerton College while working as a machinist and, later, a truck driver. The1984 sleeper hit, *The Terminator*, launched his directorial career. Since that time, Cameron has served as writer, producer, director, and/or editor on such films as *Rambo: First Blood Part II, Aliens, The Abyss, Point Break, Terminator 2: Judgment Day, True Lies,* and *Titanic*. Cameron's films have also earned numerous nominations and awards from a variety of organizations, culminating in *Titanic's* 11 Academy Awards, including Cameron's three Oscars for Best Picture, Best Direction, and Best Editing. In 1995, Cameron made 12 dives to the *Titanic* in preparation for his feature film. Cameron has made a total of 38 dives in the Mir submersibles. His most recent expedition to the hydrothermal vents is the subject of the IMAX film, *Aliens of the Deep*.

In twenty years of directing stunts and action and pyrotechnic effects, I've never actually had a serious injury on the set. That is because of application of a fairly rigorous and disciplined process. It is not as institutionalized as it is with NASA, but it has its own special rigor. Before any major stunt, or gag as we call them, I would walk the set myself, looking at every piece of rigging and turning over every possibility in my head. At that moment on a shoot when all the lights and the cameras are set up, it is the culmination of months of planning, engineering, testing, and rigging. The industry's leading experts up to that point have done it all. (I would just like to point out that the failure of the Genesis spacecraft was not due to the Hollywood stunt pilot. Of course, now we know that the science has been recovered, so it's all good.)

These experts have decades and decades of experience doing stunts, explosions, car gags, fire, and whatever it is that we might be doing. But still, even after every single one of these people has signed off, I walk the set. I just call a complete hold. I walk the set. I look at the rigging. I ask questions. I think about it: What if this happens? What if that happens? Even though we have been over it and over it, I call that last minute hold, and I walk the set. I'm looking for something which is something that I've over the years come to call the x-factor, some previously unseen detail or some exotic combination of variables which could cause the stunt to go horribly wrong.

I guess my point here is that the personal touch is critical, and taking individual responsibility is critical, for everybody in the chain. Systems protocols and institutional checks and balances are important, and they add great robustness to risky operations. However, those very checks and balances can often inhibit individuals from speaking up or taking action because they make the assumption that someone else has approved it. Someone else is going to catch it. Someone else has responsibility, and they don't catch it before it's too late.

[When] we made the movie *Titanic*, we began that production in a very unusual way. We actually dove to the wreck site of *Titanic* twelve times. It's in 12,500 feet of water in the North Atlantic. We set ourselves some pretty ambitious goals. We were going to build a new camera system so that we could operate a 35mm movie camera outside the submersible, seeing ambient pressure at 5,500 psi. We were going to build new lighting equipment. We were even going to build our own remotely-operated vehicles so we could explore the *Titanic* wreck internally. I had some experience as a project manager developing new technology for underwater filming on the movie *The Abyss*, and that prepared me, to a certain extent, for the difficulty of engineering this new equipment. Nothing prepared me for the chaos introduced when we took that whole circus to sea on a research ship. We weathered three hurricanes and multiple equipment failures, but we managed to prevail and get the images of the wreck. In that process, I got bitten by the deep ocean exploration bug.

After the success of *Titanic*, the movie, I found myself less interested in Hollywood filmmaking and more interested in the challenges of ocean photography and exploration. So, over the next few years, we developed new images and robotic exploration technology. Then I had to go out and raise the money by making films, in order to pay for it. So we wound up returning to the *Titanic* wreck site in 2001, because I figured if I couldn't raise money to go to the *Titanic* wreck I couldn't raise money to do anything. We took our spanking new 3-D digital imaging system to capture the coolest stereo images of the wreck that we could before it disintegrates, and we made a film called *Ghosts of the Abyss*, which was for the IMAX 3-D theaters.

We also created two very tiny and advanced ROVs [Remotely Operated Vehicles] which could fly untethered inside the wreck. They were untethered in terms of a power umbilical, but they had a data tether, which was a spool of fiber optic, kind of like a wire-guided torpedo. We were able to explore the wreck, room by room and deck by deck. These were launched from the submersibles after we landed on the *Titanic* wreck, and they were flown inside the wreck by myself and two other pilots. With these bots, we were able to capture some pretty amazing images inside the wreck in spaces which could never have been seen by human eyes and which probably will never be seen directly by human eyes. We were able to reveal in the lights and video cameras of these tiny robots a kind of lost grandeur of *Titanic*, which still exists deep inside that wreck.

TITANIC HAS A VALUE AS A KIND OF PARABLE. THE LESSONS LEARNED ARE STILL VALUABLE FOR US IN OUR CONTINUING EXPLORATION OF THE SEA AND OF SPACE.

For me, that was the greatest adventure imaginable. If I wasn't hooked before, I was certainly hooked then. Of course, all the time I was very cognizant of the risks and, as the person heading the team, the expedition leader, so to speak, it was my responsibility. The buck stopped with me, so I was continuing to apply my lessons learned from my underwater motion picture filmmaking experience, to this new realm. Of course, we had a lot of problems, and we had equipment failures, and we got hit by another three hurricanes. Then the September 11th attacks cut short our expedition. It was certainly a bizarre and ironic experience to be, literally, down at the bottom of the ocean, at the site of the defining disaster of the first part of the 20th century, while probably the defining disaster of the first part of the 21st century was taking place over our heads without our knowledge.

Having made 24 dives at this point by the end of the second expedition to explore the *Titanic*, I am now pretty continuously mindful of the lessons of *Titanic* as I continue with other exploration projects and any projects involving risk of any kind. The lessons learned from the sinking of *Titanic* caused sweeping reform of the maritime safety code in its time. But in the abstract, *Titanic* has a value as a kind of parable. The lessons learned are still valuable for us in our continuing exploration of the sea and of space.

Titanic was sunk primarily by institutional momentum. Just as the inertia of the ship was too great for the crew to be able to turn it in time to avoid hitting the iceberg, the inertia of their methodology was at least equally responsible for the collision. It was the policy of sea captains at that time to maintain full speed until they'd spotted the ice and then slow down only when it became absolutely necessary. This was for economic reasons, reasons of straight commerce. This was simply how it was done.

THERE ARE A FEW INTERESTING PARALLELS BETWEEN THE SINKING OF THE *TITANIC* AND THE LOSS OF THE *COLUMBIA* SPACE SHUTTLE AND HER CREW. IN BOTH CASES, THERE WERE UNHEEDED WARNINGS. IN BOTH CASES, THE WARNINGS WERE DISMISSED, NOT OUT OF NEGLIGENCE, BUT FOR REASONS THAT MADE SENSE BASED ON THE EXPERIENCE AND INSTITUTIONAL MEMORY AT THAT MOMENT.

The *Titanic's* captain was due to retire after this one last prestigious voyage, after a long and unblemished career. He was captaining on the maiden voyage of the largest vessel ever created. His lifetime of experience taught him that on a crystal clear night, in a flat calm ocean, he was safe maintaining full speed, despite the Marconi-gram sitting in his pocket warning of a huge ice field ahead. With a warning to the officer of the watch to be extra vigilant, he went to sleep as the ship barreled on toward its fate. Now, was this arrogance or hubris, as many have said? I don't think so, not really. It was simply business as usual. These new ships didn't handle like the previous ones. They took longer to stop or to turn. So, everything he knew was actually wrong in that exact circumstance. The old operating methods didn't really apply. The conditions had changed, but the methods hadn't kept up. It also required an unlikely combination of elements to create the disaster. It was a typical cascade failure where you had a number of things in series, all of which had to happen in that unique combination. The flat calm of the ocean meant that no swells were breaking against the icebergs, which reduced the ability of the lookouts to see the icebergs in the dark. The general mistake made by the crew was to underestimate the perversity of the ocean, even when it seemed at its most benign.

There are a few interesting parallels between the sinking of the *Titanic* and the loss of the *Columbia* Space Shuttle and her crew. In both cases, there were unheeded warnings. In both cases, the warnings were dismissed, not out of negligence, but for reasons that made sense based on the experience and institutional memory at that moment. In the case of *Titanic*, the crew was well aware, because of wireless messages, that the ice lay ahead, but it was because it was the way it was always done that they proceeded at full speed toward the ice field.

With *Columbia* it was known from many past missions that the foam could separate from the external tank and possibly strike the orbiter, but that problem had been analyzed twenty years earlier and dismissed as a serious threat to mission safety. When foam was observed possibly striking *Columbia* during the launch, some engineers were concerned. But because this was the way we've always done it, the warnings didn't propagate up the chain of command with enough force to change the outcome. So cultural momentum and institutional memory had worked against *Columbia* just as they had worked against *Titanic*.

Another parallel is that in both accidents an unlikely series of events were required to cause catastrophe. With *Titanic*, it was the unlikely event of the very first iceberg that they spotted, the very first one out of a huge field of ice, happening to be exactly in the track of the ship. This was occurring on a night without the slightest swell activity to assist in spotting the berg in time. And all of this was happening to a new, large class of ship whose crew was inexperienced in managing it in fast turns and sudden stops. With *Columbia*, it took the foam strike incident, but then compounded it by the fact that this was one of the very few missions in recent years that did not go to the ISS [International Space Station]. Had it been a mission to the Space Station, it is likely that the Station crew would have seen the large hole in the leading edge of the wing during the operations. Then the station could have provided safe haven for the *Columbia* crew while everybody scrambled to launch a second orbiter to bring them all home safely.

So the vanishingly small possibility of a foam strike event actually damaging a flight-critical component was coupled with the statistically low probability of a non-ISS mission to create a disastrous outcome. These low-probability, high-consequence events are the hardest to plan for and prevent, especially when it requires a number of low-probability events in combination in order to create a threatening scenario. *Titanic* teaches us to be constantly vigilant, to assume nothing about our methodology, to constantly ask the question "What are we doing wrong right now?"

I've lived with the lessons of *Titanic* and they've informed my judgment on subsequent expedition projects. After our second expedition to *Titanic*, we looked for other projects with more and greater challenges, of course. The following spring we imaged the wreck of the *Bismarck*, which is 16,000 feet down in the North Atlantic. Then, we followed that up with stereo imaging at five hydrothermal vent sites along the mid-Atlantic ridge. We were pretty excited by the imaging results from that, and I decided to make a second IMAX 3-D film about the life surrounding that hydrothermal vents. It was my intention with this film to draw a kind of sea/space connection, on the basis of a kind of ocean analogue, where we would bring NASA experts in analogue missions and let them draw the parallels between undersea operations with multiple vehicles deploying robotics. The submersibles would be like a Mars surface rover on a traverse being deployed from a habitat or a base camp for which the support ships or the surface ships were the analogues for that. But we were also drawing a connection between the types of life that existed in these chemosynthesis-based environments down at these hydrothermal vents with the kind of life that we might encounter in extraterrestrial hydrospheres; if we were to find subsurface water on Mars, if we were to find evidence of ancient life on Mars, it might have originated in hydrothermal communities. If Mars was once covered with water or had a lot more water, that water may very well have been under ice, it may have been denied the ability to photosynthesize, it may have had to live within a chemosynthesis-based environment.

When we go to Europa, we may find evidence of life there, again, probably subsisting on a nonphotosynthesis basis. So we draw the sea/space connection

in that film. I just thought I'd point that out since it's a room full of space buffs and we're an ocean panel—that there's a message there.

In the process of making this film, I've formed a partnership to buy and operate two deep rover submersibles, which are actually codesigned by Sylvia Earle. They have a depth rating of 1,000 meters. They are wonderful subs. You sit inside an acrylic sphere and you feel like you're in a kind of vacuole within the ocean where you have unlimited visibility. You feel much better contact with the environment than you do looking through the small port windows of a typical deep submersible.

Now, previously, we'd been working with the Russian Academy of Sciences and with their Mir submersible operation, which is a two-sub operation. I had a pretty good understanding of the working systems of those subs and of how the submersibles were operated and how two subs are operated in tandem with each other in diving ops. But I was certainly in for a very rude awakening by just how difficult it is to operate a manned submersible system when you're starting from scratch and when you're the one in front of whom the buck stops.

Now, we began by assembling a new team to operate and maintain the rovers and these were gathered from established submersible operations around the U.S. and Canada. The first task was to tear the subs down to their frames for ABS [American Bureau of Shipping] certification. Then we had to make all the modifications to adapt our 3-D digital technology with the [pan and fill] systems and our special lighting and all of that to the submersibles. That was certainly a daunting task. It took about six months, and we were barely re-certified in time for our filming operations.

So, we took our beautiful new subs to sea and met with the Russians out in the middle of the ocean, literally rendezvoused two ships in the middle of the ocean at the mid-Atlantic ridge. We dove them together in a joint diving operation with the Mir submersibles. This culminated with one dive where we actually had four subs rendezvous at the bottom in 870 meters of water at a site called "Lost City," which is a low temperature hydrothermal vent structure, a very interesting place.

This operation was very complex logistically because it involved the creation of new operational protocols for the launch and recovery of four subs in the same theater of operations at the same time. Tracking, communications, surface ship operations, the number of submersibles in the water made all these significantly more complicated. This had ramifications through every kind of contingency you could imagine. It wasn't just twice as complicated as operating two submersibles—it was some multiple. There was some square law at work there. Also, we were dealing with underwater communications, which, if any of you have done this, you know that such communications can be spotty at best, and we were dealing with them in Russian and English between four vehicles at the same time on the same frequency.

So, we had to go through a pretty rigorous process of defining our comms protocols before the fact. It was only because we had a good, long, healthy working relationship with the Russians that made that possible. I found that the principles of risk management and safety assurance that I learned as a film director were actually transferable to these new situations, at least at an abstract level, and certainly at a motivational level for myself, in terms of applying the same kind of energy and passion to the safety of the operation as to the aesthetic results of the film making. Now, obviously there's a very extensive body of established procedure for submersible operations, and we studied that pretty rigorously, and we selected our team members accordingly on the basis of their experience with manned submersible ops. But it seemed like almost everything that we were doing was unprecedented, and it was often difficult to find any kind of existing guidelines in the literature. Often, we were making up our own protocols in terms of what the safe procedures were for the launch of multiple subs or the manner in which we could descend them together for imaging purposes—sometimes only a couple of meters apart, how we could operate them on the bottom (proximity operations), how our acoustic comms would work during the dive, how we would work on the bottom with four subs together and a deployed ROV in the same area—a tethered vehicle.

We were able to pull on our experience from past dives, and we were able to anticipate and talk through in advance most of the contingencies that might arise on the dive. Because of the complexity of our dive ops, we always preceded each dive with a joint dive ops meeting between the Russian group and the American group. I call it the American group, but it was really a mixed group of people from Australia, Canada, and everything else. The Russians called us the "American group."

We used models of the submersibles to talk through the maneuvers. The process there was very simple. Here is a model of your sub. You are the pilot of the sub. You move it. I will give you a voice command. You move that model the way you think what I am telling you to do should result in action. This worked very well. So, literally, it would be, "Hands off. Okay, I want you to do this. I want you to do that." If they couldn't visualize it on dry land where we could hear each other perfectly, then it certainly wasn't going to be any better down at the bottom. Until we figured out what message for voice communications was going to foul us up on the bottom, we wouldn't go into the water. That was one technique.

We had perfected that in doing our wreck dives where we had the wreck as the central focusing element for what we were doing. It got more complicated when we went to these vent sites, and we were unable to physically model the vent sites. We had, in some cases, good microbathymetry, and, at the very least, we had some decent site maps. We would use those as guides, and people would fly their models. We always knew in advance what we were trying to accomplish. This briefing would then get synthesized into a dive plan document, which was distributed to all of the various crew members. You have to appreciate that we had two observers and a pilot in each Mir, so that's six. We would have an observer and a pilot in each of the deep rovers, so an additional four. Ten people were all going in[to] the water, all having to know exactly what they were doing on a daily basis. An interesting lesson here was that the task loading from a planning standpoint became greater than the task load on the actual dive. In fact, I wound up getting most of my sleep during descent and ascent because I was spending the night working through the documentation for the dive the following day. The pace of operations was inappropriate to the scale of the logistics of what we were doing. That was the thing that emerged.

Each crew member got a dive plan which was individually tailored to their vehicle in terms of the timeline and their activities—the individual objectives for each crew and the science activities as well. The science activities required a separate pre-dive meeting by the science group who would bring us their requests and recommendations for modifications to the sampling equipment on the front of the subs.

One of the things I would like to express here today is the idea that, regardless of how much you plan, you have to be willing to accept the idea of failure. I think that we are enthusiastic fans of exploration, probably everybody in the room, but failure is a part of exploration. It is absolutely woven into the fabric of the act of exploration. By definition, exploration means you're doing something that has never been done before. It is absurd to assume that activities without precedent can be done in complete safety. If only the remote and hostile environments are yet to be explored, then we are inherently pushing the limits of human endurance and technical adaptation every time we advance the boundary of what is known.

It is absolutely important to use all of our accumulated knowledge to be as safe as possible. However, safety is not the most important thing. I know this sounds like heresy, but it is a truth that must be embraced in order to do exploration. The most important thing is to actually go. Because if safety were the most important criterion, we would not go to Mars for 10,000 years, because only then could we assure absolute, 100 percent success. Historically the success of cultures and nations has been the result of their ability to balance risk and reward—to put it another way, caution and boldness.

The problem with exploration is not the individual's perception of risk; it's the institutional, national, and political perception of risk. Astronauts are smart people—I know a few of them. Most of them are Ph.D.s in one thing or another—engineering, physics, medicine. They know that riding a pointy end of a metal object that is screaming through the atmosphere at 20 times the speed of a rifle bullet, being propelled by one long continuous explosion is not quite the same as sitting at home in your Barcalounger.

They understand the dangers. They get it. They have assessed the risk. But their personal dream, their vision—not for themselves, but for the entire

. . . SAFETY IS NOT THE MOST IMPORTANT THING. I KNOW THIS SOUNDS

LIKE HERESY, BUT IT IS A TRUTH THAT MUST BE EMBRACED IN ORDER TO

DO EXPLORATION. THE MOST IMPORTANT THING IS TO ACTUALLY GO.

human race—dwarfs that risk. They know the importance of what they are doing, because in their souls they are explorers. It's not the astronauts who are going to hold up the progress of exploration. It's the government that funds them, and the people that empower that government to act, who will set the limitations. Institutions gravitate inexorably toward a value system in which any risk becomes unacceptable, at which point exploration ceases.

Now, we are lucky right now to be on a cusp with history where a presidential mandate has put NASA back on track with a renewed vision for exploration. NASA has reorganized around the guiding principle of exploration beyond Earth's orbit. This is all very exciting, it is all very new, and it is definitely happening. I believe it is a wise plan, and an affordable and achievable plan.

But there is one huge challenge that still needs to be overcome, even if we deal with all of these short-term reorganization issues. We must overcome the fear of failure that may inhibit future leaders from allowing these missions to proceed. The challenge will be this: the only way to fail in landing humans on Mars is to actually go. If we study the problem, we build tools and systems and so on for the next 50 years, we can kind of jolly ourselves along that we are really, honest-to-God going to do it someday, that we're still those clever Americans who put a man on the Moon back—when was that again?

That way we don't put our self-image at risk. But the second the button gets pushed and we are really going, then we enter a much higher realm of risk.

"Failure is not an option" was a good credo for getting the Apollo 13 astronauts back home safely, but as a driving principle, it doesn't really work. Failure must always be an option, or we stop being an exploring species.

When I started our most recent expedition project, I called a big summit meeting of all the department heads. I stood in front of a white board and put up on the white board three slogans. The slogans were there: "Luck is not a factor," "Hope is not a strategy," "Fear is not an option." Now, the first two were meant to convey my philosophy that to succeed in any complex task, it is essential to leave nothing to chance. You need to make your own luck by rigorous application of a robust process. You test everything in a very disciplined fashion, you don't guess, you know the answer, you anticipate every negative condition that might possibly prevail. You assume it is going to happen. You have an A plan, a B plan, a C plan, and you assume that you're going to be on the C plan by your second cup on coffee on morning one of the expedition, because that's how it goes when you're at sea.

I wanted to scare them, and I wanted them to respect their adversary—not the ocean, but the real adversary: entropy, which, as you know, is the tendency of things to go from a state of organization to a state of chaos.

The third slogan, "Fear is not an option," was meant to inspire the boldness that actually sees you through these endeavors. It was the yin and the yang of the healthy paranoia which the first two slogans represented, because without a kind of faith, which is not in luck and not in passive hope, but in yourself and your team and in the greater meaning of what you're setting out to do, you won't find the strength to go through with it.

So my message is, in whichever realm, be it going into space or going into the deep sea, you have to balance the yin and yang of caution and boldness, risk aversion and risk taking, fear and fearlessness. No great accomplishment takes place, whether it be a movie or a deep ocean expedition or a space mission, without a kind of dynamic equipoise between the two. Luck is not a factor. Hope is not a strategy. Fear is not an option.

Over the Edge of the World

Ferdinand Magellan took almost three years to circumnavigate the globe. In fact, he didn't make it. He was killed in the middle. Jules Verne wrote about going around the world in 80 days. I am going to take you around the world in Magellan's tracks in about 10 minutes, much more safely than Magellan did. In terms of risk and reward evaluation, keep in mind that, of the approximately 260 sailors in five very small ships that he took, leaving from Seville, Spain in 1519, only one ship with 18 sailors made it back three years later to Seville. One ship mutinied in the Strait of Magellan and returned early. Over 200 hundred sailors died in this attempt to circumnavigate the globe. That was not exceptional.

In this era of exploration, in the 16th century, it was a different mind-set. The very rational and logical and useful tools for evaluating risks and rewards didn't exist. The mind-set was closer to the medieval mind-set, even though this was the quintessential Renaissance exploration mission more than anything else. We can see, despite that mind-set, modern tools and paradigms and approaches emerging. Nevertheless, people went with an expectation that if they succeeded, it would be God's will, and if they failed, that was God's will. That was Magellan's inspiration for going, and that turned out to be, as you'll

Laurence Bergreen
Author

Laurence Bergreen is a prize-winning nonfiction writer. His new book, *Over the Edge of the World: Magellan's Terrifying Circumnavigation of the Globe*, was published to great acclaim by William Morrow/ HarperCollins in October 2004. His last book was *Voyage to Mars: NASA's Search for Life Beyond Earth*, a narrative of NASA's exploration of Mars and the search for extraterrestrial life, published in 2000 by Penguin Putnam. He has written for *Esquire, Newsweek, TV Guide, Details, Prologue,* and *Military History Quarterly,* and many other publications. A graduate of Harvard, he has taught at the New School for Social Research and served as assistant to the president of the Museum of Television and Radio in New York. In 1995 he served as a nonfiction judge for the National Book Awards and in 1991 as a judge for the PEN/Albrand Nonfiction Award. He also serves as a featured historian for the History Channel and is a member of PEN American Center and the Authors Guild.

see when I get to the part about Magellan's death, his undoing as well, despite his many crew members urging him to ignore what he felt was God's will.

His mission to circumnavigate the globe for the first time ever was not meant to be a scientific one. That concept really didn't exist. He was going for two reasons, and they were pretty basic reasons. One was for greed, and one was for glory.

There were two superpowers in those days, one of the important analogies to the recent present, during the Cold War. Those superpowers were Spain and Portugal, and they were vying for control of the ocean. They were doing that because they were vying for control of the world economy, or the global economy, such as it was. The key to that economy in those days wasn't oil, the way it is now; it was spices. We all say, "What's the big deal about spices—cloves, cinnamon, nutmeg? Who cares? You can buy them in the supermarket." In those days, cloves were the most valuable commodity on the face of the Earth. They were more precious, pound for pound, than gold.

On his voyage, Magellan refused a number of opportunities to trade iron for gold on a pound for pound basis, because he wanted to save space on his ship for the cloves, which were more valuable. That one surviving small ship, less than 90 feet in length, *Victoria*, that made it back to Seville laden with cloves, made enough money for the bankers who financed it and for King Charles, the Spanish banker, to make the whole expedition—which was, in human terms, a tremendous disaster—a huge commercial success. This inspired Spain to follow up five times, each time unsuccessfully, on Magellan's vision of circumnavigating the globe.

For me, researching this book, there are two approaches. One is the library. People often say, "Well, where did you go to research this book?" And I usually quickly deflate the balloon by saying, "To the library," because that was the most important place. However, the library really isn't enough. You really have to get out into the field. It always reminds me of when I was a kid and dropping those little paper Japanese flowers into water—just add water and they come to life.

When you go to the Strait of Magellan or you go to Sanlucar de Barrameda, the port city in Spain from which Magellan's ships left, you begin to see the scale and the scope of what it was like. When you walk across a life-size replica of one of Magellan's ships and see how tiny it was and how primitive it was, you realize that what they were taking looks to us, on a temporary risk-reward evaluation basis, to be doomed to failure. But they didn't think that in those days. They thought that God was going to be on their side. And I'll try and explain a little bit to you why.

So for me, this was mostly tourism, to go in Magellan's tracks. Someday, to go in the tracks of Neil Armstrong or Jim Lovell will be mostly tourism. Not yet. And it [would have] seemed inconceivable 500 years ago that tourists would go through the Strait of Magellan the way I did, with a couple of friends with our cameras, walking over glaciers that [had] imperiled Magellan's life and the lives of all his sailors. And the glory part of this was that they were going to bring Christianity and the glory of King Charles—who was all of 18 years old when he commissioned Magellan to go on this voyage—around the world and spread those two goals far and wide as they went to claim the Spice Islands, which are in Indonesia, for Spain. Nobody really knew exactly where they were, and part of the expedition would be to find a shortcut, a fabled route somewhere through the South American landmasses to the Spice Islands. The exact size and shape of South America was not really known. Something was known about the eastern coast and that was all. They would cross what was known as the Pacific Gulf, considered to be a very small body of water.

My book on Magellan actually began on Mars with my previous book, *Voyage to Mars*, which was about NASA's robotic exploration of the red planet through four missions, from Pathfinder through the ill-fated Mars Polar Lander. During that time, NASA scientists at Goddard and JPL [Jet Propulsion Laboratory] kept talking about precedents for their exploration of the universe. They kept talking about Columbus; we all know a lot about Columbus. They talked about Balboa. They talked about Vasco da Gama and they talked about Magellan. And, after about the tenth or maybe the twentieth time, the name Ferdinand Magellan

SO FOR ME, THIS WAS MOSTLY TOURISM, TO GO IN MAGELLAN'S TRACKS. SOMEDAY, TO

GO IN THE TRACKS OF NEIL ARMSTRONG OR JIM LOVELL WILL BE MOSTLY TOURISM.

was mentioned to me, a dim light bulb eventually illuminated in my mind, and I thought that might be a very interesting idea for a book. It might have a lot to say about our own current age of exploration of the solar system and the universe. Because after all, at the time that this man, Ferdinand Magellan, went around the world, the world was as mysterious to Europeans as the solar system and the universe is to us.

Who was Magellan? First of all, he was a misfit. If he was on this panel today, he probably would be the least popular member. He would be the one that everybody would be looking at and saying, "He looks like a fanatic. He looks like a weirdo." He wouldn't have that genial, easy-going manner and that self-deprecatory humor that we admire in pilots and captains who are undertaking high-risk missions. From the little bit that we know from contemporary observation, he had a knack for being abrasive and for offending people. He defected from Portugal, because he couldn't get backing from the king of Portugal, who personally disliked him, to Spain, where he really wasn't a known quantity. He was preceded by a reputation as a daredevil, Portuguese soldier and a mariner, but he was an unknown quantity. And he quickly managed, through some sleight of hand, to get backing from the king of Spain and his backers, who were older and wiser, because they were desperate to beat Portugal to the Spice Islands, much as this country was desperate to beat the Soviet Union to the Moon during the height of the space race.

Magellan was limited by his communication skills—he never learned Spanish well. He was very embarrassed by his heavy Portuguese accent. He usually communicated through very stiff documents. If he ever cracked a joke in his life, there was no record of it.

But he was an obsessive personality and two things obsessed him. [One thing was] navigation, and he was a perfectionist in navigation. And as a Portuguese, he was aware of what was then the state-of-the-art of navigation and cartography in the world. Portuguese were like the Soviets in the space race, obsessively secretive about their cartography. If you published a book in Portugal that contained any map or information about Portuguese voyages, you were thrown into prison. And, of course, the book was destroyed. This was, of course, after the age of Gutenberg and [the publication of] Columbus's books had been a very important way of disseminating information. In fact, Columbus was Magellan's boyhood hero, and when Magellan read Columbus' account of his first voyage to the new world, that inspired him to go even further than Columbus, the way some astronauts today are inspired by their childhood memories of watching John Glenn and other astronauts, and their exploits, on television.

So Magellan, putting it mildly, was not a people person, but he was a brilliant navigator. He was also obsessed with one other element of his fleet of five ships, which were all leased and were all in bad condition: food. Most of the records that we have of that time—and they are voluminous—show that he was exceedingly careful about provisions and feeding the men what he thought would be the most effective diet. And tremendous thought and care was given to the kind of food, even though it was all horrible food, it was all salty. It was salt beef, it was salt cod, it was salt pork, there were olives. The only sweet thing was honey, which was taken along, and there was a tremendous amount of wine, which was the staple beverage. It was mixed with water, so it probably wasn't very tasty. And the other staple element was hardtack, that was basically stale biscuits. It was a month old by the time it even got on the ships, and it gradually became wormy and rotten and soggy as the voyage went on. And even when it was soaked with the feces and urine of the rats which infested the ships, the sailors continued to eat it because there was nothing else to eat beyond their rations, except for the leather wrapping the masts of some of the ships.

You may wonder why anybody would want to go on a voyage like this. In fact, most of the sailors came from the convict or semiconvict class and had no other hope for their survival in Spain but this voyage of escape from whatever their current problem was. Perhaps it was marital problems, perhaps it was debts, perhaps it was some crime that they had been accused of and this was their one escape. The officers were motivated often by greed because, after all, if they could bring or smuggle back some of these cloves, they would be set for life. Even a sack full would be enough for them to purchase a small house in the sailor's suburb of Seville and live there comfortably for the rest of their lives.

Magellan went because he believed that he was going to discover a new world. He really was impelled by what we would call idealistic motives. Now, the king of Spain gave him tremendous latitude. He was given the ability to name continents and islands after himself, none of which he did; in fact, he turned out to be very self-effacing. The Strait of Magellan, for example, he named the Cape of the Feast of the 11,000 Virgins, which doesn't really roll off the tongue that well, because that was the feast day on which he discovered it. So he was giving primarily religious names to places he discovered because he was a very devout individual.

His crewmembers came from at least ten countries. They spoke at least ten languages, and they didn't get along. They consisted of a number of cabals, and the Spaniards didn't talk to the Portuguese, who didn't talk to the English, who didn't talk to the Germans, who didn't talk to the Norwegians, who didn't talk to the Greeks. You may wonder how they communicated just to get ordinary sailing and nautical tasks done. They used an argot that was a Catalan slang that they all understood. But there was no easy rapport among these crewmembers, who would just as soon get into fights with each other as cooperate on their missions.

I think it's fair to say that Magellan, with his lack of so-called "people skills," faced much greater obstacles from the individuals on board the ship and the people he encountered in their travels around the world than he did from natural obstacles. In fact, he learned to master most of the incredibly overwhelming natural obstacles, including terrible storms in traversing the Strait of Magellan, which is a nautical nightmare. But he never really knew how to handle people, except with the most brutal means imaginable, such as torture, in order to inspire and put dread in the men to follow him.

The major player at that time was King Charles the V, the king of Spain and the emperor of the Holy Roman Empire. The king of Spain was a Hapsburg king; you can tell that by that famous Hapsburg jaw in the portrait of him by Titian. And it was in the name of King Charles that Magellan went. Keep in mind King Charles was an 18-year-old boy; he was trying to grow a beard when he sent Magellan on this mission, and even when the survivors came back three years later, he was only 21 years old and widely mistrusted by everyone around him.

The other major player in that era was Pope Leo X, who, as reflected in the portrait by Rafael, was a worried man. And if those Cardinals that are around him look like they are menacing him, it's because, in those days, the Cardinals were routinely plotting to kill and poison and strangle each other, and there were constant plots against the life of the Pope. Nevertheless, Magellan went around the world constantly pledging his loyalty and his entire expedition to the greater glory of the Roman Catholic Church and bringing the Church enlightenment to people around the world.

Magellan did not bring slaves or try to enslave people, which was a big difference between him and his boyhood hero Columbus. He did bring one personal slave with him, but when he found the so-called heathen in places, his first thought was not like Columbus's, "Aha! There are so many people here we can enslave!" His first thought was, "Aha! There are so many people here that we can baptize." Also, the group aboard included men, women, and children. So this already marked a very important shift from the previous era of exploration.

The maps they used at that time were worse than useless. The so-called "T&O map," based on a literal interpretation of the Bible, shows the ocean circling the world, only three continents, Jerusalem at the top—it was absolutely useless for anything. And this was the way most people—although not scholars—looked at the world at the time [of] Magellan['s travels]. This would be circa 1515.

There was a state-of-the-art map that was based on the calculations of Ptolemy, the famous Greek-Egyptian mathematician whose mathematical compilations were rediscovered and published during the Renaissance. The map is a projection of the world as a sphere, based on his calculations. The one dramatic omission is the Pacific Ocean—9,000 miles! Had Magellan known that after he accomplished his greatest feat of navigating the Strait of Magellan, which is at the southernmost tip of South America, that he still had to cross the Pacific, he probably wouldn't have gone and he probably wouldn't have gotten backing from the Spanish crown or from the financiers who were expecting—like businessmen everywhere—a reward, a return on their investment. They were hoping for about 14 percent, incidentally. So his maps were mostly useless. In fact, when he got to the Pacific Ocean, he was so exasperated with these kinds of maps and charts that he threw them overboard in a temper tantrum and said, "These maps are not to be trusted." And from then on, he relied solely on his own charts.

A map of his actual route gives you an idea of how it looked in the world as it actually was. And you can see his route as he leaves Seville and goes to the coast of South America following a well-worn path by that point, until he begins to work his way down to near the southernmost tip, looking for the Strait of Magellan, which he had promised his backers and the king of Spain he'd find or else. And, finally, he did manage to find it. But from then on, for most of his route and for most of those three years, he was sailing through waters that were uncharted by European cartographers and so were unknown to him.

We don't have an image of his actual vessel. But we know it rides very high in the water. It's dark brown or black because of the pitch, the tar covering the sides to keep it seaworthy. Depictions of ships of the time are shown surrounded by flying fish, which were a constant fact of life of some of the earlier parts of the voyage, and by some sort of sea monsters, which were believed to exist.

Some of the hazards that they believed to exist at that time were mermaids, considered to be a fact. Another was a magnetic island; if the ship sailed too close, the island would pull all the nails out of the ship, the planks would come apart, end of story. That was also considered to be scientific, factual. Then there was the mythical continent of Terra Australis. Not until the 19th century was the existence of this continent, thought to somehow counterbalance the continents of the Northern Hemisphere, disproved. And, also, the water was thought to boil at the equator, because it would be so hot.

So, there were all sorts of imaginary hazards that Magellan and his sailors thought they were facing, which turned out not to be the case. However, they

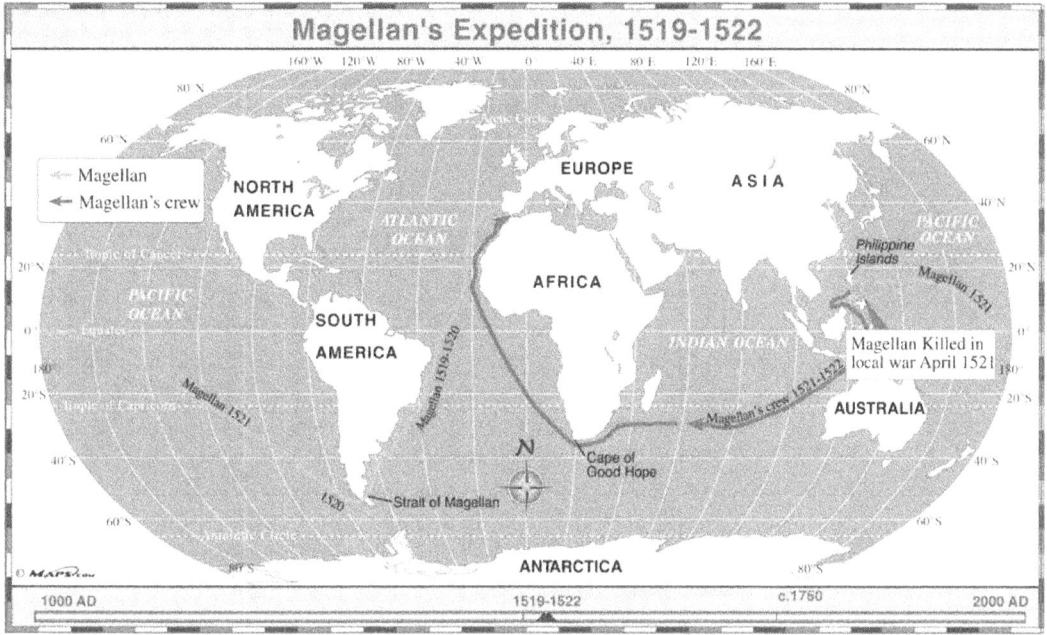

Magellan's Expedition, 1519-1522

were also facing real hazards that were in some ways even more dangerous. For example, scurvy. Scurvy was the radiation poisoning of its era. There was no known cure for scurvy. We now know that a teaspoon or less of Vitamin C taken a day, in orange juice, or many herbs, or even beer, or malt, is the magic bullet cure for scurvy. But Europeans didn't know about that until 200 years after Magellan's voyage, which was a complicated, fascinating medical story in itself. At that time, scurvy was a dread disease. It caused loosening of the teeth and mottled skin, and then, literally, the hard parts of your body, your bones, your teeth, your tendons, would come apart—your body literally falling apart. And over thirty sailors on Magellan's crew succumbed to the deprivations of scurvy and a horrible death at sea because of this disease, which we now know is so simple to prevent.

So, danger was everywhere, and again, prayer and a belief in the divine will was about the only protection that the men felt they had against it.

One of the great false leads of the voyage was the Rio de la Plata, South America, which many of the men insisted was actually the Strait of Magellan. Of course, it's many hundreds of miles north. When Magellan saw it was shallow and covered with silt, he figured, just based on sheer instinct, that it wasn't deep enough to somehow cut through the South American land mass, and come out the other side in the Pacific Ocean. And so he sailed around the bay, and kept going, and said, "This is not it." The men didn't agree, and they mutinied, and he responded to the mutiny by drawing and quartering some of the leaders, which was a brutal procedure that involved removing their intestines while they were alive, burning them in front of them, and, eventually, decapitating them, putting their heads on a stake, and putting those stakes in the harbor where the ships were moored in order to enforce discipline. And that was how Magellan kept his men in line. It was a very, very different era from today, as I was saying at the outset.

One of his chief discoveries, which was truly accidental, was the Magellanic Clouds. Now, as I mentioned before, this was not a scientific mission. Not until the Age of Enlightenment and Captain Cook in the 18th century was the concept of a science mission really popular or prevalent. Nevertheless, Magellan brought with him a chronicler who had never been to sea named, Antonio Pigafetta, a funny name. But Pigafetta was a rather intelligent, very ambitious young Venetian diplomat who heard about Magellan's voyage when he had gone to Seville, and he signed up for this mission. And most of the important things we know about it comes from the diary that Pigafetta kept on the voyage. He survived, and Magellan didn't.

These Magellanic Clouds, which were until about 10 years ago thought to be the galaxies closest to the earth, were simply described by Pigafetta as two "clouds of mist." Period. They had really very little idea of what they were looking at, but he noticed everything. He also wrote down 30 different languages that Europeans didn't know about, spoken by various preliterate tribes around the world, giving us our first lexicographies of all these languages. So he was an astronomer, an ethnographer, and an anthropologist. He also became, because he learned these languages, translator for this mission. Pigafetta turned out to be one of Magellan's best hires, let's put it that way. Especially because he survived.

The Strait of Magellan itself is unchanged from 500 years ago. It's basically a fjord. The water is very cold. If any of the sailors had fallen overboard, they would have survived six minutes at most. By the way, most of the sailors then didn't know how to swim, and they had a terrible phobia about the water.

Five hundred years later we walked across what our guides like to call a "cold beach." There were Magellanic penguins, which were ubiquitous [and] which bailed Magellan's sailors out of starvation time and again when they went through 500 years ago.

The way Magellan managed to navigate what was really a maritime maze— not a straight watery path—was to have his men climb mountains, and look ahead and see, well, which way to go. What was a dead end, and what was going to take them to the Pacific during this 300-mile crossing? He also tasted the seawater. When it was salty, he knew he was near the Atlantic. When it got to be fresher, he figured he must be getting to the middle of the strait. And when it turned salty again, he figured he must be coming out to the Pacific, which was a misnomer, because the water there was even rougher than it had been in the Atlantic, where he had faced some terrible storms.

Glaciers were noted by Pigafetta, and looked at by all the men, but they couldn't figure out why they were blue. Of course, they're blue for the same reason that water is blue, because of the way the eye selectively absorbs scattered light. Magellan's fleet was very lucky not to have been crushed in one of the glaciers.

Magellan very unwisely immersed himself in a tribal war when he reached the Philippine Islands. When he reached that archipelago, he was actually worshipped, literally, by the islanders there, whom he converted to Christianity in mass baptism, including men, women, and the children. Pigafetta calculated several thousand conversions. And Magellan got so caught up in this that he wanted to keep on doing it. Meanwhile, all his officers who had survived to this point said, basically, "You know, we're on a commercial mission here. We have to get to the Spice Islands. You don't know where they are. We've got to get there." But Magellan said he wanted to stay.

There was one island leader we know by the name of Lapu-Lapu, who was in a war with all the other islands in the Philippines. He was the ruler of the island of Mactan. And he decided that since all the others were converting to this new and strange deity who was brought from afar in these gigantic black ships, he was going to do the opposite. So, he challenged Magellan to a battle. Magellan, as you might gather, was never one to back down from a fight. And he said "Fine, we'll undertake this battle." He figured he had gunpowder and weapons, guns on his side, which were very primitive and as likely to blow up as to fire correctly. But they did have crossbows, which were far more lethal, and they also had armor. Magellan figured that armor would be impervious to blows from bamboo swords and that one of his soldiers would be able to defeat 50 or even 100 island warriors.

So Lapu-Lapu challenged him to a battle. Magellan decided that 60 men would be enough for him to handle whatever Lapu-Lapu threw at him. He waved off assistance from a local sultan who offered all of his soldiers and troops to Magellan in favor of Magellan's support. He waved off offers of support from all his men. He told his ships to stay way back—he didn't need to be covered by fire because God was going to protect him.

So, he undertook this amphibious landing early in the day on April 27, 1521. His 60 men were met by 1,500 enraged soldiers with fire-hardened, poison-tipped swords and with bamboo shields who charged into the water and eventually overwhelmed Magellan, once they figured out who Magellan was—and he was rather conspicuous because of his plumed conquistador's helmet. (Note to other explorers, don't wear a conquistador's helmet while fighting the enemy!) They managed to throw spears at the exposed parts of his body, at his arms and his legs. Finally, they managed to knock his sword out of his right hand. When he stooped to pick it up from the water, he took another spear in his arm, disabling it. And then Lapu-Lapu's soldiers closed in for the kill. And, essentially, they hacked Magellan to pieces right there in Mactan harbor, and there was nothing large enough left for even a proper burial. And that was the death, the very, very unnecessary death of perhaps the greatest explorer of the entire Renaissance era.

His crew had seen this coming, because they had been aware of his growing recklessness, and they quickly elected two captains, Portuguese and Spanish, to continue the expedition all the way to the Spice Islands. And then, finally, overcoming one disaster after another, one ship made it back.

137

By the way, the one ship that made it back was captained by a Basque mariner, Juan Sebastián Elcano. And in Spain this is known as the Elcano mission, rather than the Magellan mission, out of a nationalistic feeling, because Magellan was Portuguese and was viewed with so much suspicion by the Spanish authorities.

So, as you can see, the idea of what exploring was like in those days was almost incomprehensible compared to what we're used to today. And so our exploration of the solar system continues in that spirit, but with a tremendously different approach from what it was like then.

Discussion

DAVID LONGNECKER: My name is David Longnecker, from the University of Pennsylvania, and I'm addressing my question to Mike Gernhardt. I was intrigued by your calculation of acceptable risk for DCS (decompression sickness). As you know, the concept of acceptable risk is one that's getting considerable play throughout NASA, as we look towards future exploration. Do you think it's possible to apply such techniques and such mathematical approaches to broader risk categories, as we look for broader missions?

MICHAEL GERNHARDT: That's a great question, and the answer is: absolutely, you can. There are limitations, obviously, to the data and to the statistics, and, ultimately, you will have to make judgments. I found that running this process, I started out with an uninformed group and my own notions of what acceptable risk were. And we ended up with an educated group and a specific definition and a consensus to move forward. So, I think the answer to that question is yes, a similar process could be run with return to flight, using models that are out there for [foam-shedding] and MMOD (Micrometeoroid/Orbital Debris) prospectively defining what acceptable risk is for reentry. The only danger is that that should not be a substitute for good judgments. And I think if you look at that as one tool, a decision support tool, it could be very valuable in that regard.

EUGENE RODDENBERRY: Hello. Eugene Roddenberry. Actually, I've got a question for Mr. Cousteau. I wonder if you could tell us about what your son is doing today and if he's okay. From what I hear, speaking about risk, he's taking some risk right now.

JEAN-MICHEL COUSTEAU: Well, I think he's taking a very calculated risk. He, and some of the Hollywood people have helped him, built—that was a dream of his for a long time—a life-size great white shark in which he's hiding. The structure of the great white, which can move on its own, is such that anything can attack it and he's completely safe. What can go wrong is his life support system if he doesn't do the right thing. So, it comes back to him. It's not nature that's the problem, it's human.

His objective is to find himself in the middle of other great white sharks, perfectly protected, a lot better than I was when I was in South Africa. And he, from the inside, can see through the eyes of the shark, as can the cameras which are looking out through the eyes, at what's going on around it, and can kind of study the behavior of these sharks, which we know so very little about. So, from a scientific point of view, hopefully, they will make some new discoveries as to the behavior of great whites by including less risk than if we were in cages or even scuba diving. So, I don't think he's taking a lot of risk, personally. Much less than other people have, and, hopefully, we will learn something. And that's what he's doing at the moment, as we speak.

DAVID LAWRENCE: David Lawrence for Laurence Bergreen. Was going around the world the great challenge in 1519 that Magellan hoped to meet, or was it just to get the cloves and get home, and it was just accidental that his expedition continued to make the first circumnavigation?

LAURENCE BERGREEN: Yes, that's a good question. The latter. It was almost incidental. He figured that was the fastest way to get to the Spice Islands to bring home the spices and to avoid the time-honored overland route, which was much slower, far more expensive, and controlled by the Arabs. So, it was really what he felt was the expedient way to do it. The efficient way to do it.

GORDON OSINSKI: Gordon Osinski, from the University of Arizona and soon moving to the Canadian Space Agency. We've talked so much, so far, about the risk of exploration, and the title of the symposium, but until this morning, nothing about the risk of not exploring. I think John Chatterton said, "Exploration is who we are. We should continue on the path of exploration or quit." And Sylvia Earle said, "Something is happening to us as a species." There are people yesterday who thought, to continue evolving as a species we should explore, we have to explore.

I was moved by the reasons why we should explore the sea, and we're looking to the stars. I was born a few years after man last walked on the Moon. I've been doing some teaching recently and I'm shocked, aghast, at how many people think we have not walked on the Moon, or actually, how many people think we have walked on Mars. So my question is: Is there a greater risk of not exploring than exploring? And maybe pose that to the whole panel and to everyone.

JEAN-MICHEL COUSTEAU: I would just like to jump in by saying that we've done a very, very poor job of communicating the results of our exploration to the public. I mean, you'd be amazed to go in parts of the country and find out that people don't know anything about what's going on at NASA. And we have to see a communication resolution, that we are leaving at the moment and taking for granted. We have to find a way to get, particularly, young people to know what's going on. And by doing so, we're going to revive the excitement of exploration and stop, once and for all, this concept that everything has been done, and everything has been discovered, and there is nothing to do, and let's go and have a drink. It's very, very sad, and I see this more and more. But there are people who are starting to make a difference in that sense. So, we need to really tackle young people in schools.

SYLVIA EARLE: If I could jump in on this. I was so dismayed by this cover story on a new magazine, I think it was *U.S. News and World Report* last spring, about the great age of exploration being over. That the great frontiers were during the time of Magellan and Columbus and all that. And yes, there's much to be done out in space, but this planet is largely explored. I actually sat in an Explorer's Club banquet one evening and listened to a spokesperson for space exploration talk about how the only frontiers left were up in the sky, until Kathy Sullivan kicked the speaker from under the table, and I got up and gave him a laser look from across the room. We're talking Carl Sagan here.

And he backtracked and said, "Oh, yes, most of the ocean has yet to really be explored." And that's the point, you know? If I were in charge, the administrator of an agency with the objective of looking at the solar system and surveying all the planets and all the things and even beyond, I'd say, "That blue one! That one there with all the water. That's the one we really need to concentrate on because that's where the action is!" If you're looking for life, find the water. And we've got it. It's here.

And my greatest fear is that we, with all of our technology and knowledge about how dependent we are on the natural systems that support us, we're going to let the system degrade to the point where our species is going to be in trouble. We are in trouble! The thing is, we don't appreciate it. I'm all for looking skyward and in every direction of exploration, but it baffles me why we aren't really motivated to look inwards. To look at the ocean, to explore it, and to find a place for ourselves here within the natural systems that sustain us. And to apply this great technology that we have to really understand the magnitude of what we don't know about the ocean, and put it to work for us, for our survival, for our well-being. This is the time.

In the next ten years, if we don't really take action, we're going to lose the chance with many of the species that we have taken for granted all our lives—tuna, swordfish, and the like. It's going to be gone! Coral reefs and all these other systems that are at risk right now. We have the capacity to turn things around. The real question is, are we going to use our knowledge in the spirit of exploration to do it? I mean, Goethe said, "It's not enough just to know. You must act." Well, we know. Do we have the capacity now to act?

JOHN CHATTERTON: The spirit of exploration is certainly one thing. But exploring requires resources. It requires money. And right now, we're very much satisfied with spending money on weapons of war, on SUVs, on things that are really counterproductive to our best interests. And, certainly, one of those things would be exploration.

JAMES CAMERON: Well, I think that's an excellent point, you know. I guess I tried to make it—probably crudely—that the type of exploration that remains to be done on our planet requires more advanced technology than previously. You could do a lot and put your names in the history books with a small ship—which was state-of-the-art at the time—or some sled dogs and some true grit

141

and some luck. These days, none of those things are sufficient. You need large organizations like NASA or NOAA, Wood's Hole [Oceanographic Institution] or MBARI [Monterey Bay Aquarium Research Institute] or some body like that to provide technology and support staff and engineering and so on, so it does boil down to a budget issue.

Go back to Mike's formula, you know? Cost is a factor, the likelihood of success is a factor. You run that equation. That applies to the financing [of] a movie, the funding of a deep-ocean expedition, or an entire research program that might deal with the deep. People look at it and say, "What's the reward? What's in it for me?" But there are new and interesting ways to finance explorations that didn't exist before. The deep ocean is revealing such vast biodiversity that whole new genomes are being revealed, and there are pharmaceutical companies that are interested in bioprospecting the deep ocean, which will allow them to create new drugs, new treatments, and so on. So there's renewed interest in pure exploration, in a sense, and biosampling in realms that previously were being overlooked as not economically viable.

So it's just a question of being creative about how we create the funding paradigms. I've tried to do something a little bit unusual. In the past, filmmakers have piggybacked on scientific expeditions that were going anyway for reasons of the goals of their various parent institutions. We flipped it around on our last film and got the money from the media sources, then went to the scientific community and said, "Hey, we're going out with submersibles to the hydrothermal vents in the East Pacific Rise and the Mid-Atlantic Ridge. Who wants to come along and take advantage of these assets that we're marshaling out there for imaging purposes?"

And interestingly, our best response was from the astrobiology community. We wound up taking researchers from Ames and Johnson Space Center and from Jet Propulsion Laboratory with us out there to—not to do analogue research, but to look at the biology of the deep vents and relate it to what they might find in the fossil record on Mars or other places in the solar system and beyond. So we were actually using media and entertainment funding to help with science and exploration. There are different ways to skin the cat, but I think the important thing is for everyone collectively to try to engender the passion for exploration in the next generation.

And part of that is reminding them of the heroes of the past and keeping that image alive, and part of that is reminding them that there is so much of the world and of the universe that has yet to be explored. It is within our grasp, and it's a real adventure that we can really have and really enjoy in our lifetimes, if we put our will toward doing that.

SYLVIA EARLE: What is the cost of not exploring? That's the real factor.

ANDY PRESBY: My name is Andy Presby. I'm a student here at the school. When a person of my meager accomplishments attempts to suggest something that may be new to a group such as this, he must do so with a certain degree of humility.

I hope you will recognize the respect that I have for everyone in this room and everyone, particularly, at that table. I think you guys are missing the point.

I've heard a lot of talk over the last couple of days, and I've been a space nut since I was three years old and my daddy took me to see one of Mr. Roddenberry's movies. However, we talk a lot about the scientific benefit. We talk a lot about the personal exploratory benefit. For example, exploring Mount Everest. You then go and talk about the need to inspire the next generation of explorers, and I completely agree with you, Mr. Cameron, that that is absolutely required. You've talked about the cost. We talked yesterday about a goal, and typically, when I hear somebody in the space community talk about a goal, they mean a planet. They mean a body. I don't think that's the goal that will inspire that [next] generation of explorers.

We have for the first time in human history come to recognize that, as you say, Mr. Lovell, we are living on a spacecraft, a giant spacecraft that we didn't design and we don't know how it works. Now we're screwing around with the life support mechanisms. I work in submarines. I understand and fully appreciate the need to keep life support gear running, but I also understand the need to explore the environment around me and, perhaps, find alternative means through which the needs of—what is it now?—six and a half billion people who all want the American standard of living, which, if I am not mistaken, involves approximately two personal slaves worth of energy per year per person.

We live in an environment that's flooded with energy. Space is full of it. I hear folks talking about very narrow goals, and we're talking about scientific goals. We're talking about exploratory goals and personal goals, and those are all important. Those are all immediate short-term goals, but I argue that, to inspire the next generation of explorers, you need to speak about long-term goals such as finding ways to relieve the pressure that we place on our environment by looking at, and yes, it's going to be expensive, sir, ways of moving resource production and other systems such as that off planet.

I have a tremendous interest, and everyone sitting here in this row has a tremendous interest, in the sources of human conflict. The two sources of human conflict, as my friends have said, are the misunderstood "other," close proximity to same, and lack of resources. Why don't we speak about that?

DAVID HALPERN: Very well said by the next generation.

JAMES CAMERON: I wholeheartedly agree that energy is probably going to be the source of conflict. It is currently the source of conflict. It is going to continue to be the source of conflict, and there may be energy sources revealed, whether it is mining helium-3 on the Moon, doing off-planet fusion production, creating antimatter on the back side of the Moon where the earth is shielded, or whatever you want to do. I think these are good ideas. I think that the issue of solving the problems with our life support system here—which is something I personally am passionate about but didn't speak about today because, frankly, I knew my colleagues here would do it because I know them well—is a separate issue from exploration. I think that there are aspects of exploration that are survival requirements.

143

When Sylvia talks about the risk of not exploring, it's really the risk of not having the technical capability to explore. We build our muscles slowly to go out and do these things. We are still on an indefinite hold in low Earth orbit, building up the muscle to learn how to support human beings for long duration in order to be able to go further. We are relying on our robots now to be our precursors out there. We have to build up this capability, and, personally, I believe you do have to have goals to do that. You have to have a focusing element. It can't just be an abstract thing: Let's go out and solve our energy problems out in the universe. We won't solve them at the Moon. We'll solve them here, generating the technology that enables that exploration.

In my mind, I uncouple the abstract goal of exploration, which is to satisfy the human soul—yearning, understanding, all those things—from the hard core nuts and bolts activity of exploration, which has always spawned so much in the way of economic enhancement of this country and of the other developed countries, because we put so much energy into the technology required to do these difficult and exotic things. We will develop an improved nuclear power system. We will develop fusion power. We will develop some of these things, and the control systems for same, in the course of trying to get to Mars or do these high energy things that we have to do in order to explore the solar system. Our understanding of distant stars from orbiting next-generation space telescopes and so on may be the key turning the latch of figuring out how to have an unlimited power supply here on Earth that will replace oil, and give us another different excuse in the future for going to war. Right now, our excuse is oil.

SYLVIA EARLE: While we do look for alternatives to our current energy sources—and we should definitely do that—meanwhile, we can make better use of what we've already got: more efficient use of our current energy resources. It is not just in terms of oil, gas, and things of this nature; I mean in terms of food resources, too. Twenty million tons of wildlife extracted out of the ocean is simply thrown-away bycatch. More than 300 thousand marine mammals every year are destroyed in the process of catching fish. We are seeing the fish that we are taking just collapsing. You know we are too good at catching these things. We are hunter-gatherers, but we're armed with new technologies that our predecessors could not imagine. So, we need to put on the brakes and think about more effective use of the resources that are here. We couldn't support six billion people with wildlife from the land. Ed Wilson, Harvard biologist, says we've seen consumed "the large, the slow and the tasty from North America over 10 thousand years". It's only taken us 50 years with our new technologies to do the same thing with the ocean. We are very close to losing some of the creatures that we have thought infinitely able to rebound no matter [how many] we extracted from the ocean. Exploration, in terms of finding solutions to the very problem you have posed—how do we find the place for ourselves that is going to last, knowing that our numbers have increased three times in my lifetime, but the planet stays the same size? Our capacity to support us is currently being stretched. It's not just oil and gas. It's oxygen in the atmosphere. What are we

144

doing to that part of the world, the ocean, that is generating most of the oxygen, absorbing much of the carbon dioxide? We're messing around with it. We need to know how it works. That means explore it, and then, take heed. Not just, yeah, we've got all this new information, but acting on what we are learning and doing it in a way that secures a place for ourselves so we can continue to explore as long as humankind survives.

ANDY PRESBY: I don't know if anybody else wants to comment, but I didn't mean to focus specifically on energy. It was an example, and one that we can all relate to. I don't know if that helps anybody respond to my question.

JIM GARVIN: Jim Garvin, NASA, Moon and Mars Chief Scientist. They're our resources. I think the tenet I'm hearing in response to this great question is that we have to separate exploration, as a catalytic tool to make things better, from the applied end game of exploration that we can document in history, from Magellan's search for cloves and in finding first orbit of Earth. How do we measure that? One of [the] things we are asked all the time is, what is the yield from these catalytic things? Whether they be to inspire, what are they? We use lots of terms, and I think this audience would be wonderful to try come up with those metrics. This young man says inspiration isn't enough. Okay. As we catalyze, what is? The one I always find easy, maybe because I'm simpleminded and not yet quantum-computing, is IT. Information technology. Why are we doing it better in some places? Many reasons. Smart people. Maybe that's an area we ought to look at as part of exploration to extend ourselves to think better and to use our resources better to better inspire. Anyway, that's my comment for the group.

JAMES CAMERON: I think there is an inspirational dividend to exploration. I think this is one of the primary reasons to do it. I think you have to ask yourself, why are the Chinese doing a space program that basically mirrors what we were doing thirty years ago? Why is it important to them now, as the fastest growing economy on the planet, to be doing it, to simply be reproducing an accomplishment that is already done? Because they know that the inspirational dividend within their own borders is going to be significant in inspiring kids to go into technical careers in math, engineering, and science. So, the value that they are getting out of it is much greater than what they are putting into it. They've done the math. They can't win that race any more.

I think we should ask ourselves, what are we losing by not exploring, in terms of the inspirational dividend to a younger generation? One of the biggest problems this country is going to be facing is the lack of "fresh outs" in engineering, math, and sciences in the next ten years or so. We've got some big problems to solve, and we're too far down the path as a technological species to go back to the garden and try to pretend none of this ever happened. We've got to get ourselves out. We've got to think ourselves out of it as a technological species. We have to continue to build those tools and that capability.

Certainly, with the vast amounts available for military procurement, you've got people working in math and science and so on making pretty good livings there.

145

Wouldn't you rather have an alternative to that, though, in space exploration where we can focus our minds, improve our IT capability, improve our control over energy systems, and, by the way, understand long-term regenerative life support systems? If we're going to go to Mars, the point is to stay there, not simply touch base and run back. We're going to have to learn how to live there with very, very finite resources.

The more we learn about closed loop ecosystems, the more we need to know about the big, closed loop ecosystem that we live in, and vice versa. One body of knowledge will feed the other. I think there is an awful lot to be gained societally for the investment that we are making in space exploration, and I would certainly like to see a proportionate amount spent on ocean exploration. I know Sylvia is enlightened in that she is one of the few people in the ocean community that doesn't constantly complain about those billions that NASA gets.

When they're fighting for tens of thousands or hundreds of thousand of dollars, she is enlightened enough to know that all knowledge improves us all and that exploration should be constantly going in both directions. I agree with that as well. I could talk for hours about the value of analogues, of ocean exploration, of space exploration . . . And how you could build muscle in both places . . .

JOHN CHATTERTON: The other thing is that exploration should not just inspire more exploration. Exploration should inspire additional exploration, but it should inspire us to think big, to work on problems like energy, to work on problems like the environment, to work on problems like population. Everything that we've got on our to do list as a species, we need to apply ourselves, if we're going to find solutions. We don't really have that much in the way of a choice.

JIM PAWELCZYK: My name is Jim Pawelczyk and I'm at Penn State University. James Cameron, you mentioned in your talk that we've already picked the low-hanging fruit with regard to exploration. All of you have spoken about inspiring the next generation of explorers. Do we need a different educational paradigm in order to make those things mesh? And if so, what do you think it looks like?

SYLVIA EARLE: I have three children and four grandsons. It disturbs me that we aren't getting this generation coming along—the kids—actually out doing things in good, wild places. In fact, our safety mechanisms in schools dictate against it. You go to Hawaii, kids aren't allowed to go get in the water as a part of their school activity—not above their ankles, anyway—because, you know, it's not safe. So whatever it takes, whether it's museums, aquariums, moms and dads, whoever, we need to take the responsibility for getting kids connected with the real world, the living world, the wild world. We're missing it in the rather structured form of education as it is currently being conducted, not just in this country, but most of the rest of the world where education systems are—really, I mean, it's important to learn the ABCs and the 1-2-3s, but we've also got to learn that we're a part of this greater system, and that's missing.

PAUL SPUDIS: Yes, Paul Spudis, Applied Physics Lab. I want to thank Andy for stimulating a really good conversation here, because I think he's nibbling around the edges of something. I've been listening to this for the last couple of days and I've heard a lot of interesting things. But I have two comments. First, in regard to this argument of spending money on weapons versus exploration, they're actually complimentary and, in fact, historically, the exploration is something we let the military do during peacetime. And all the great explorations of the Pacific weren't undertaken because they were interested in the natural history of Polynesia, they actually wanted good maps that they could use to retain British control of the seaways.

The second point I want to make is that I think that we've nibbled around the edges of the issue of why we do exploration. And I think there's three motivations to it, of which we've only discussed two. The first motivation was discussed yesterday, and that's sort of the personal gratification. You know, because it's there, I want to go, I'm curious, I want to know. The second motivation is societal and collective. It's, we explore to get strategic information, to inform ourselves so that we can make better guesses on how to do something else, whether it's to identify other resources or to develop a technology or something like that. But no one's talked about the third motivation. And that is exploration as a prelude to settlement. We explore because we want to go live there. And one of the really interesting things that we got out of Apollo is an appreciation for the fact that, sooner or later, life on this planet is doomed. We know this because we know that impacts occur, and we know that in the past they've come darned close to nearly completely sterilizing the Earth—wiping out almost 95 percent of all living species. So, ultimately, someday, somehow, that's going to happen here.

And one of the big motivations, I think, for exploring space, is to create additional reservoirs of human culture, so that if Earth is destroyed, or the biosphere is destroyed, there will be, the human race will survive. Now, that's a long-term thing, certainly isn't a part of going to the Moon or going to Mars. But doing that by going to these places, we're going to learn the skills we need to develop the ability to live off-planet. Does anyone have any comment on that?

PENNY BOSTON: Penny Boston, from New Mexico Tech. I've been thinking about another type of risk that we really haven't addressed yesterday and today. And it's really a risk to exploration. When I look at everybody here who's doing exploration, we're all relying more and more on ever greater degrees of technology. So that the point at which one can participate in this, the number of people becomes narrower and narrower. You have to be well-educated and you have to have access to resources. And I think back to the famous essay by C.P. Snow, in the middle of the last century, the two cultures where you see this increasing dichotomy between those who know and those who have, and those who do not know and do not have. And it seems that, unless we attend to that growing bridge in society, that ultimately threatens our future in terms of exploration. I see symptoms of that in these sort of vacuous reality shows that are on TV. As much

147

as we may denigrate them [the reality TV shows], what it seems to me to indicate is that the vast majority of people are feeling more and more uncomfortable with their excessively cloistered and safe lives, and that, perhaps, this need to acquire risk is general throughout the population, even though people like us manifest it maybe more obviously. And so, this potential danger to all of our enterprise, whether it be ocean or land or space, seems to me a festering element that we need to address.

JAMES CAMERON: I think that's an excellent point. And my answer would relate to the previous question about education. You suggested that the problem is that the technology narrows the band of people that can actually participate. But, in fact, technology can also be an enabler for people everywhere to participate, through improvements in information technology. And, you know, theoretically, we're all wired up to one big human nervous system. So, if we have an avatar, whether it's robotic or human, out there somewhere at the bottom of the ocean or in space doing something that's interesting, there's no reason why we can't all look over its shoulder and participate. But it requires a will on the part of the people budgeting that operation to make sure that they put in as a line item, not just outreach in the sense of, "We're going to tell people what we're doing and show them some images," but participatory outreach in the sense that, "We're going to let you look over their shoulders. We're going to spend that extra two or three percent on a major mission to let people actually participate." And I know that the recent activities on Mars have done an absolutely stellar job in doing that, if it can be judged by the number of hits to the NASA Web site—I think it's up to 11 billion now or something like that. People are looking over the shoulder of those little rovers. And if we had human beings there right now doing microbiology—I know that's your field—or whatever, if we should get so lucky as to find some evidence of that on Mars, people would be able to participate in that. So I think the solution is always going to be there as a technical solution. It's a question of imagining it before the fact and incorporating it into what we're doing.

SYLVIA EARLE: Just endorsing your observation about the need to have risk. It's a kind of spice. Probably more valuable than cloves.

DAVID HALPERN: With that parting comment, I think it seems appropriate to bring this session to an end. I'd like to thank Administrator Sean O'Keefe and Ames Director Scott Hubbard for the wonderful facilities that we're in now. And I especially want to thank each of the panelists for their dedication and their wonderful comments and their inspiration for what we've been doing. And I'd like to thank, finally, the audience for the wonderful questions and wonderful attentiveness. And with that, I have one more thing. Those of you who are from the East Coast probably have never lived through an earthquake, but you just had a 5.9 earthquake about 120 miles off the coast. So this here session is memorable in many ways. Thank you again.

stars

Introduction

In this symposium, we've heard from women and men exploring the earth, the seas, and the limits of human capability. We've discussed personal risk, programmatic risk, technical risk, survival of the species risk, and the most important of all, the risk of not exploring at all. We are compelled by some ancient instinct to push the limits, to go where humans can't survive except for brief periods of time or with significant technical support.

My name is John Grunsfeld. I'm the chief scientist of NASA. I'm an astronaut; I've had the privilege to fly four times in space. I've done five space walks, so in fact—along the lines of going to places where people can't survive—I do, in fact, work in a vacuum, along with many of the others in this room. And it's truly a privilege to have been able to be involved at the infancy of space exploration. In this session, we do turn to that ultimate challenge, our first steps off the home planet. We live in a truly remarkable time.

As we speak here, as we're comfortably sitting in this environment, Gennady Padalka, the commander on the International Space Station, and Mike Fincke, the chief science officer, are spending their 163rd day in space. (I may be off by a day.) For over three years, we've had 24/7−365 occupation of the International Space Station. The ultimate service, if you will. And I think that's pretty remarkable.

Spirit and opportunity are still alive on Mars; we'll hear more about that. Cassini is at Saturn taking unbelievable images, things that we've never seen before, things that we certainly don't understand. We have the Hubble Space Telescope, the Spitzer Telescope, and the Chandra X-ray Observatory astronomy facility—three of the four great observatories, all exploring space and discovering things that we couldn't even imagine when these instruments were conceived.

We now know that there are over a 100 planets around nearby stars, when a decade ago, we only knew about our own solar system. And, in spite of this tremendous growth of our knowledge of our home planet, the solar system, and the universe, it turns out from recent observations that we only know a tiny bit about what makes up our universe. Ninety-six percent of the universe is filled with stuff and we don't have a clue what it is.

But I have to say, we're a little bit arrogant. Because when I was a graduate student and a postdoc and a faculty member at Caltech, it was believed that we knew about most of the universe, the history of the universe, the Big Bang and inflation and expansion of the universe. And I thought that most of the great frontiers in physics had been solved and we were cleaning up the details. And, as you heard this morning on the sea, we talk about having explored to the ends of the Earth, but 96 percent is still unexplored. For most of the universe, we still don't have a clue what makes it tick. We are really in the infancy in space exploration. Only 12 people have walked on the Moon, our closest planetary surface, and it's time to leave the cradle.

In all the preceding talks, the central theme has been risk—that's what we're here to talk about. One element that has been discussed, peripherally or centrally, is what I consider to be one of the central issues, which is teamwork. Space exploration sets a new extreme as a team activity. I think we can draw a parallel to Jim Lovell on Apollo 13, the ultimate team of folks 200,000 miles from planet Earth with a pretty terrible problem. And Mike Foale, as he described last night, on the *Mir* space station, an international team also with really tough times. And how the team in space and on planet Earth came together to solve those problems that led to the ultimate success of those missions, the safe return of the crew.

For this session, we've assembled a team to continue this great discussion on risk. And I dare say we have, indeed, put together a team of stars.

John Grunsfeld, NASA Chief Scientist and Astronaut

Risk and Reward in Apollo

One of the things that occurred to me that might be worth emphasizing, relative to other discussions that have already occurred, is a brief summary of the reward that came with whatever risk that was run—personal as well as national—with respect to the Apollo program. Certainly it was conceived in the context of the Cold War, and it succeeded spectacularly. And even some of the émigrés that I've had an opportunity to talk with say that it had a tremendous influence on the confidence of the Soviet leadership relative to President Reagan's Strategic Defense Initiative, in that they believed—maybe more so than many of the people in this country—that we would succeed, because we succeeded in Apollo where they had not.

Secondly, the technology base that Apollo enhanced—I think you're hard put to find any specific item that it created, but it certainly enhanced the technology base. That technology base is available to us still today and has accelerated human progress in so many different ways and in so many different fields.

The cultural and societal legacy is often, I think, forgotten. It was a tremendous confidence-builder among the American people at the time. And, really, as we traveled the

Harrison "Jack" Schmitt
Former NASA Astronaut

Harrison "Jack" Schmitt is a geologist, astronaut, and U.S. senator. He received a bachelor of science degree in science from the California Institute of Technology in 1957 and a doctorate of philosophy in geology from Harvard University in 1964. He was a member of the Apollo 17 mission in 1972 and, thereby, the first and only geologist to land on the Moon. Schmitt served from 1977 to 1983 as a United States Senator (Republican) representing New Mexico. Harrison is a founder and chairman of Interlune-Intermars Initiative, Inc., an organization that advocates the private sector's role in lunar exploration and developing lunar resources. He is chair emeritus of The Annapolis Center (risk assessment evaluation) and an adjunct professor at the University of Wisconsin-Madison, teaching "Resources from Space." Harrison consults, speaks, and writes on policy issues of space and the science of the Moon.

world as ambassadors during the Apollo 17 postflight tour, it was a confidence builder for people all over the world. Now, whether we've lived up to the legacy of that confidence or not is another discussion, but nevertheless, these kinds of projects do have that kind of effect, in that, if we can go to the Moon, then we surely can do some other things. And the answer is that you can if you motivate young men and women to believe it's the most important thing they're going to do with their lives. If you can create that kind of motivation, indeed, you can do just about anything.

And, finally, from my perspective of actually having been on the Apollo 17 mission, the scientific legacy is just unfathomable. It is absolutely a magnificent legacy of Apollo and its precursors that they created our modern understanding of the origin and evolution of the Moon, a foundation that's been built upon by some of you with the Clementine and Lunar Prospector missions. It's something that now relates directly to further understanding of the terrestrial planets, not the least of which are the Earth and Mars. Imagine, once in a while, what it would be if we did not have that legacy of information about the Moon—what kind of thoughts you would be having on Mars based on the information currently

coming in? So this, in a very brief way, I think illustrates why the reward was so fantastically important and so much worth the risk that a few people, and the Nation and managers and families, took in pursuit of that goal.

What I would like to spend a little more time on today is thinking about the probability of success, which is the inverse handmaiden, if you will, as a measure of risk. A few years ago, in a paper that I'm sure nobody saw that I gave at one of the space conferences in Albuquerque, I tried to deal with the evaluation, in a semiquantitative way, of the various approaches that one might take to return to deep space, and, specifically, to the Moon. As you might expect from my biases, the private sector approach won in that evaluation, but it was on the assumption that there are commercially viable lunar resources—namely, as was mentioned by James Cameron and stimulated by the young man here to my right, the possibility of lunar helium-3 fusion power as one of several potential sources of electrical power that we're going to need over the next 50 years and beyond. If you want the latest lay analysis of what the envelope of financial and technical success is for a lunar helium-3 initiative, I will recommend to you the October [2004] *Popular Mechanics*.

But, nonetheless, clean, low-cost energy is one way that we can solve many of the problems discussed this morning. It is clearly the challenge of this generation and subsequent generations as well. My own estimate is that in order to just provide four-fifths of the world's population with the level of standard of living that we enjoy today, we're looking at 10 to 11 times the amount of energy by 2050 that we consume today per capita. I'm not going to go into that any more deeply, but if someone is interested, I'm sure I could find the paper that would go into it more deeply.

Well, ladies and gentlemen, risk is always with us, as has been made very clear by this outstanding symposium. And there are always going to be people around us, many of them in this room, that are willing to take the risks, whatever they might be, because we can conceive of the rewards. We're human beings, and one of the great advantages we have as a species is we conceive of these kinds of things. In major technological-based endeavors, I have come to the conclusion,

. . . OUR EXPERIENCE WITH APOLLO AND SUBSEQUENT ACTIVITIES INDICATE

THAT YOU NEED TO HAVE MOTIVATING OBJECTIVES THAT ARE ABOUT 10 YEARS

APART, PLUS OR MINUS A COUPLE OF YEARS.

studying this over the last 20 or 30 years to some degree, that there really are three dominant, interrelated determinants for success.

We were talking about the probability of success here, which is the inverse of risk. One is the size of the management reserve funding; second is the management experience and flexibility to carry out this great project; and third is a cadre, a reservoir, of motivated young men and women.

Now, first let me address the size of the management reserve of funding. In Apollo, you all know the story—true or not, but it certainly worked out that way—of Jim Webb getting an estimate of what it was going to take to accomplish Apollo and then doubling it and doubling it again, and that was our management reserve. And we used all of it. We didn't go over it, but we used all of it. And the reason a management reserve [is] so important is [that] it enables management, then, to deal with the unknown unknowns and with erroneous initial assumptions that might have been made about the approach to the problem.

Apollo is a good example of where an adequate management reserve brought success. The Shuttle is an example, particularly in the early days, of where an inadequate management reserve caused significant problems. I happened to be in the Senate when Shuttle was headed for its first flight, and in 1978 it was clear that Shuttle was not going to get to first flight without a major influx of funds. The Carter administration, at the time, was not willing to fight for that supplemental budget. I think it was the fiscal '79 budget, if I remember correctly, that needed the supplemental. Until a good friend of ours named Hans Mark was able to push and persuade President Carter that the Shuttle was required for verification of the SALT II treaty, which Carter was very interested in, we were not going to get that supplemental.

But, fortunately, Hans did persuade the President to do that and we suddenly had the White House helping us get the supplemental through a Democratic Congress. I might say that Hans had set that one up a little bit, because the payloads had been designed for the Shuttle and not for ELV [Expendable Launch Vehicle], so he had a pretty good argument. That does present, though, an ethical dilemma, and I don't have an answer for this dilemma. It depends on what you believe the ultimate value and reward of the project in hand will be, in that, if you, in a legitimate analysis, know that you have inadequate funding and continue with inadequate management reserve to take on a new project, should you take it on? Maybe the Postgraduate school here could have a seminar or something on that subject and try to come up with an answer, but I really don't have it. Because I think you can argue that the managers of NASA at the time felt that even if they realized they had an inadequate management reserve of funding, that if they didn't go ahead with the Space Shuttle under the constraints that the Nixon administration established, that we would not have a manned space flight program. So I'll leave that dilemma with you, and maybe you can debate it over the next beer.

Now the second thing, the management experience and flexibility, was certainly epitomized by the NASA management team in place between 1967 and, in particular, in 1968–72, when it really did crystallize and became a team that is to be envied, I think, by all of us. And it was critical to have that kind of team, based initially from the heritage of the NACA, the National Advisory Committee on Aeronautics, and added to by people like Sam Phillips and Bob Seamans and others, who came in and provided that really remarkable team that led us through to success, not only for the first landing, but, also, the success for Apollo as a whole. And, indeed, that team of managers, in particular Bob Gilruth, George Low, and Sam Phillips, made decisions well prior to Apollo 11 in order to optimize the Apollo system so that it could be used for scientific exploration of the Moon. All you'd have to do is look at the decisions that were made. The lunar rover, the Block II Lunar Module, the advanced ALSEP [Apollo Lunar Surface Experiments Package] and, strangely enough, the agreement that we should begin to fully train the Apollo crew, beginning with Jim Lovell's crew, in exploration geology. And the combination of all of that meant that we got that legacy I mentioned earlier of a remarkable scientific return from Apollo.

The third and maybe most important determinant for success is to have that reservoir of young men and women available to apply their stamina and their imagination to the project at hand. Jim Lovell did not mention it yesterday, but Gene Kranz, after the Apollo 13 crisis had been resolved, did an analysis of the average age of people in the Mission Control Center and it was 26 years old. And most of them had already been there for several years. And there was just no question that you've got to have that kind of stamina, that kind of imagination and flexibility and willingness to work as teams, or you probably are not going to be successful.

Now, having said that, I also feel that our experience with Apollo and subsequent activities indicate that you need to have motivating objectives that are about 10 years apart, plus or minus a couple of years. But that's about what I think human evolution has given us in terms of our ability to concentrate for 16-hour days, 8-day weeks for a long period of time in order to make sure that which is to be done happens. So, as we look to establishing a long-term capability for indefinite exploration of our solar system and, eventually beyond, I think we have to still think in terms of how we quantize that period into specific objectives that each generation can identify with and accomplish. Now, with respect to public support and political support, I don't think there's any question that we in the United States believe and will continue to believe—and certainly NASA believes—that visibility and transparency are absolutely essential.

If you're going to have active and sustained political and explanatory support for efforts like we are about, you also need to have a White House deeply committed and involved at basically all levels. There is no substitute in the Congress for an active interest and activity from the White House. Whether it's a Democratic or Republican-controlled Congress, you still have to have that. Otherwise, you just don't get their attention very well.

Now, on the private sector side of things, you need to have investor support. And to have investor support, the most important thing, of course, is to have competitive returns on investment. That is, competitive with other uses of capital. If you can do that, if you can show a path for that return on investment in a relatively short period of time, you have a predictable path for success. In fact, it's more predictable than the government, and that's why we're going to concentrate on that right now. I know what the criteria for success are in a private initiative. I know what they are in a government initiative. But the ones in the private sector are much more predictable. Or you just need to have an angel out there with irrational exuberance. And we're hoping to find one of those one of these days, and we'll be on our way.

Back to Contents page

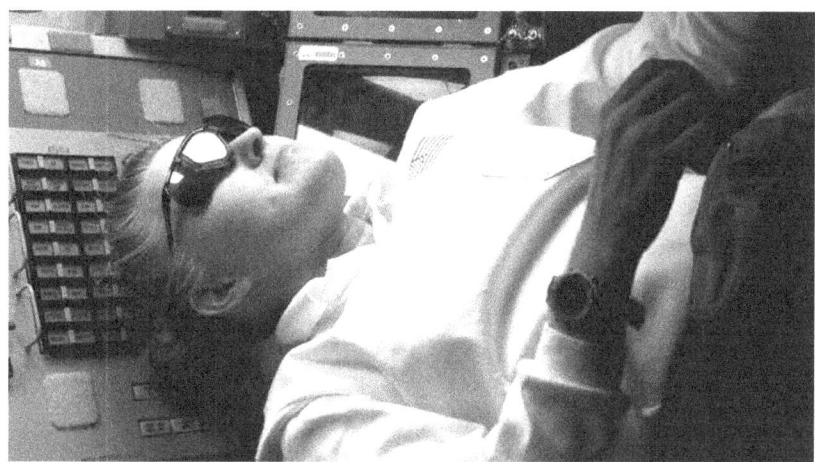

Mir Adventures

John asked me to get on the panel and talk a little bit about personal risk. How do you decide, from a very personal standpoint, what risks you're willing to take? Specifically, to talk about the thoughts that went through my head, when I agreed to be a crew member on *Mir*, in connection with personal risk. It's a fact that a person is born and from the time you are born, you are taking risks. Every single person, every day of their life, is taking risks in one form or the other, and, as you go through life, you are learning to mitigate the risk. What risks you are willing to accept, what risks you are not willing to accept in such and such a case. Well, I'm not going to go back to the day I was born to talk about how I think I learned to accept risk, but I will go back to 1962 when I was 20 years old. I just had graduated from college, I was 20 years old, I had a passion to fly, and I had no job. As you can see, that didn't correlate very well, but I knew that what I wanted more than anything in the world was an airplane that I could fly.

So my father asked me what I was planning on doing, and I said I was going to buy an airplane. And he looked at me and he said, "You've never even bought a car yet! How do you even know how to buy an airplane?" And I said, "Father, don't you worry! It's airplanes we're

Shannon Lucid
NASA Astronaut

Shannon Lucid was born in Shanghai, China, but considers Bethany, Oklahoma, to be her hometown. She graduated from Bethany High School in 1960, received a bachelor of science degree in chemistry from the University of Oklahoma in 1963, and a master of science and doctor of philosophy degrees in biochemistry from the university in 1970 and 1973, respectively. Shannon Lucid was selected as an astronaut in 1978, after spending time as a teacher and researcher. She has flown as a mission specialist in the Space Shuttle on five missions, deploying satellites, conducting scientific observations and experiments, some in a special laboratory built into the orbiter's cargo bay, and spending 188 days working on board the Russian Space Station, *Mir*. From February 2002 through September 2003, Lucid served as NASA's chief scientist at NASA Headquarters in Washington, DC. Lucid remains an active astronaut at the NASA Johnson Space Center in Houston, Texas.

talking about. Aviation." And I said, "Everybody in aviation is absolutely honest, so don't you worry. Nobody will sell me a bad airplane!"

So I didn't have to worry about how to go about buying an airplane or what to look for, et cetera. Well, I went out searching, and I found an airplane. Not a pretty airplane, but it was the only thing that I could get in the panhandle of Oklahoma, and, trust me, this is really true, it was sold to me by an aviation salesman with a glass eye. And he told me it was absolutely perfect and he would show me the logbook he had. It just had a brand new fabric job and so on. I bought the airplane, and the next month the airworthiness certificate was taken away from it, so I was grounded. I thought my life was over.

But life went on. And I did get a very, very valuable lesson. And the lesson I learned was that when you are involved in an activity, just because someone is involved in the same activity and maybe they're using the same words, it doesn't necessarily mean they have the same value system. And when you are working with a team, in the activities that you're in, it's very important to know that the people that you're investing with have the same value system, and that when you're talking about something, you're talking about the same thing.

The other lesson that I learned over the next few years, before I came to work at NASA, was I flew that airplane and I flew other airplanes for several thousand hours. I flew from Alaska down to Central America, and there were always concerns about whether or not you're going to have an accident. If you read aviation reports, it seems that most of the accidents were due, in one way or the other, to pilot error. And the biggest concern that I had was, okay, it's okay if you crash, but I didn't want it to ever be pilot error.

So I didn't want the obituaries to read, "Shannon was so stupid. Can you believe that she took off in that thunderstorm?" And, so, that sort of tempered the decisions that I made. And they are all hard decisions that you make while you're flying. Mainly, the hardest decision is not to fly. And I found that I had a very difficult time doing that, to say no, when I really wanted to go. But, like I said, over the years I learned how to do that.

So I went to work for NASA. And I was very fortunate in working for NASA, because with the Shuttle flights that I was fortunate to fly on, I found that there was a team that had the same values system. And one of the great benefits of flying and working for NASA was working with the people in Mission Control, working with the people at the Cape, because we were all part of a team, we were all working together for the same objective.

So, then the day came when I was asked if I would like to be a crew member on *Mir*. And, of course, I would have said, you didn't need to ask because I've been volunteering for years. But I said yes, and then a friend of mine went up to *Mir* before I did, on a flight to bring Norm Thagard back. And my friend had not been back on earth more than two hours when the phone rang, and my friend said, "Shannon, don't go." And I said, "Oh? Don't go where?" She said, "To *Mir*." And I said, "But why?" And she said, "Because you will be living in a mine shaft for several months, and I just can't picture you living there. So, take my advice, and don't go."

And so that was one input. Another input was, we trained in Star City in Russia, and the week before I was to go back to the United States for three weeks before I got on the Shuttle to go up to *Mir*, I was leaving a movie theater in Moscow which doubled as a church on Sunday morning. As I was leaving there, a lady came up to me who I didn't know, and she said, "Are you the American woman that is going to go up on that tin can?" And I said, "What?" And she said, "I saw on the news. Are you really going to go up and be locked up in a tin can for months with two Russians?" And I thought, well, now, that's a novel way of looking at it. It wasn't quite the way I had been looking at it, and [it] made me think why did I really want to do this, because I really wanted to do it in the worst way.

. . . THE WAY I NEGATED THAT RISK WAS, I CHANGED THE PURPOSE OF

THE FLIGHT IN MY HEAD . . . I TOOK WHAT WAS A PERCEIVED RISK AND

I CHANGED IT INTO WHAT I COULD HANDLE.

And there were several reasons. One reason was, because I had had several Shuttle flights, and I enjoyed them very much. I really enjoyed flying in space. And if you have a small piece of cake, a big one would be even better. So I figured that, since I had enjoyed the short Shuttle flights I had been on, I would really enjoy a longer duration space flight.

And the other reason why I really wanted to go was because I was curious. I was very curious to find out what it would be like to live and work in space for a long period of time. I was curious to see how the body would adapt. And I really wanted to experience that.

And the other reason was, as a child I'd always wanted to be part of an expedition. I mean, I read all the books about expeditions and going off and exploring the different places on the Earth. And, in my mind, going for a period of time on a long duration spaceflight on the space station would be the equivalent of going on an expedition. Because I had written to National Geographic, I had written to everyone I could think of, to find out how you could get on an expedition. And it turns out that, basically, you had to know somebody who had some money. And since I didn't know anyone, that option was closed. But this was a chance that I had to experience that. So, those were the reasons why I was so anxious and so eager to go and fly on the space station *Mir*.

Now, while I was on *Mir*, there were other risk factors that came in. One of them came when I was talking to my daughter on the ham radio one day. Anything that I found out of any importance about the program always came to me via friends or family members or the ham radio or whatever. And *then* we heard from the program. And my daughter said, "Hey, Mom, guess what? You're not going to come home when you think you are. You're going to be stuck up there for months." And I said, "What? What are you talking about?" My daughter worked for a contractor that worked for NASA, so she was tied into the chain of gossip. And she said, "They're having some kind of a problem, so don't pack your bags yet, because you're not coming home."

And the first time, as soon as I heard that, I thought, oh my goodness. Because I remembered back to meetings that we had had at JSC [Johnson Space Center], and we had had them for years before, talking about space station. They were the life science meetings about how you should plan a mission. And the gist of it was that you could only go in small increments. We had been in Skylab 90 days, 100 days, but we had to be very careful. We couldn't exceed that. We had to have something like 10 to 20 missions before we could go beyond 90 days. And then we were going to go up to 100 days, and then we were going to go up more. One of these very small increments, getting up to where we could spend a longer period of time.

So my first thought when my daughter told me that was [of] those meetings. And I thought, oh my goodness, in one fell swoop we're going way beyond anything they discussed in those meetings. My second thought was, get real, Shannon, let's use a little common sense here. Because there in Star City you've seen all these people walking around that have been in space for a lot longer than you're going to be up there. And there is nothing wrong with them.

And, so, common sense negated what might have been thought of as a perceived risk at that time.

Now, another risk that I felt before I launched was that I wasn't really sure what I was going to be doing when I got up on *Mir*. And I say that because we had the U.S. experiments that were coming up, the science. But we didn't know when it was going to be there, because it was delayed. And no one was just real sure when it was going to get there.

And the other thing [perceived risk] was due to the complexity of the program. I had never been able to go through the experiment that I was going to be doing end to end. So, thinking about it just before we launched, I thought, you know, I could be really being set up here, because if you go and you weren't able to do the experiments, you look, professionally, sort of bad because you weren't able to get your work done.

But, then, the way I negated that risk was, I changed the purpose of the flight in my head. I was not going up to do the science. I mean, that would have been nice, but that wasn't my personal primary purpose. I thought, the reason I'm going to go, and the reason I'm doing this flight, is because I want to see what it's like to have a long duration flight, and I want to make sure that my crew mates and I get along and that we have a great flight. And I thought, that I can handle. That's not dependent on any payload coming up, or it's not depending on having any procedures.

So I took what was a perceived risk and I changed it into what I could handle. Now the next risk that I thought of was, well, people always ask, "Were you comfortable living up there on *Mir*? How did you sleep?" And the reason why I was comfortable living on *Mir* and went to sleep every night without any problem was because Soyuz was always attached. It gave you a great deal of comfort in your heart knowing that if a problem arose, if there was a fire, a rapid depressurization or anything, you had a way home. You had a lifeboat. You could get in the Soyuz and go home. And that you had an automatic abort mode and that gave you a tremendous feeling of comfort.

Now, there was one other risk that arose, and that was, once we got in orbit, it turns out—and I didn't know this ahead of time—the Russian cosmonauts were going to do EVAs [Extravehicular Activity]. I thought, that's fine, they're going to go out and do EVAs, and I'm going to sit and watch them.

Well, about a couple of hours before they went out to do their EVAs, and actually, they were already in their EVA underwear, the commander called me over and said, "Shannon, quick, come here, I've got to train you, because this is what you're going to do while we're out". And I said, "What? I am going to be doing something?" And then he started rattling off in Russian these long lists of commands that I'm supposed to be putting into the solar panels to get them to move so the station will work properly. So, I'm losing my mind, so I say, "Whoa, whoa, wait, wait, wait a minute! You can't have me doing this, because I haven't been trained and I don't know how to do this". And he said, "Oh it's easy." And then he was rattling it off again. Then I said "Look, it may be easy for you, but I'm just an older American woman. So I need a little bit of help." And I said "I need a procedure". And he said, "Procedure? We don't have any procedures." And that was true, they didn't work off procedures very much. And I said, "I repeat, I'm an American. We work off procedures. I have to have a procedure." And then, because it was close to the EVA, he was getting a little tense, he sort of gave up. But then the engineer said, "Okay, I'll help you." So we sat there and we wrote out a list of the commands and procedures that I was to follow. And then we wrote out how I was supposed to know when it was time for me to do this.

And I mention this because I did feel under a little bit of pressure, because I wanted to do the right thing. And I knew I was under pressure because I had a lot of sweat on the back of my neck about that time. But I took the perceived risk, or took what I felt was a risk, and changed it into something I thought I could handle. I changed it into a procedure. I forced the system into accommodating what I could handle. And so that's how I handled those risks.

So, from a personal standpoint, that's how I looked at *Mir* and handled the risks that I saw. Then the big question that people always ask is, "Well, okay, why do you want to fly in space? You have these risks, so why do you want to fly in space?" I know exactly why I want to fly in space. It's sort of hard to put into words. That's why I put this picture up there, because instead of a metaphor, it's sort of like a "picturephor," except it doesn't really say it all. It's a sunset. And you say, "Okay, you want to fly in space because you want to see a sunset?" Yeah. It's because you can look at the world in a new and a different way. You grow, and it's a huge challenge. And one of the aspects that I really enjoyed about spaceflight is because you're working with a team, you're working with a marvelous team.

Now, being actually in space, being the person that actually goes to fly, you're the person that people see, and sort of the tip of the iceberg that sticks up. But it's this huge team effort that does it, and being part of a team, it enlarges yourself, so that you're bigger than yourself. And it's all of that put together that is the reason why I want to fly in space.

And a further reason is sort of the same reason why I really enjoyed working in a lab, back in the days when I used to be a scientist. You'd work in the lab and you'd work all hours, and then, finally, one day it would happen where you had an idea; you had done the experiment, and then you looked at the data, you saw the data, and you thought, wow! This is something new. You had found a new way of looking at the universe. Then you write your paper, and no one else thinks it's as marvelous as you thought. But, still, you had the feeling that you were able to find something new that hadn't been seen before. You'd seen the world in a new way, and it's sort of the same way when you can express in words in such a way that someone else can understand what you're saying, because you put the word combination in a certain way so that you'd gotten the message across.

And that is all part of the same reason of why you want to go into space, and why you want to explore. It's because you are part of something that's bigger than yourself, and you can get a feel for what it's like to be really creative, and really see the world and experience the world in a new and marvelous way.

NASA Image /Renee Bouchard

Back to Contents page

Mars Exploration

I'd like to share with you a very personal set of connected anecdotes to talk about the risk of exploration, in particular of Mars. We all explore in different ways and sometimes we don't know that we're exploring. I think in spirit we explore. In the 1960s, I didn't know the risks I was taking as black disks were hurled at me at 100 miles an hour, and as, in the cacophony of the hockey game, I weighed those risks. And as a young child, I experienced those as well. But I think the message in all of exploration is learning, as one explores, to accommodate the risks, to recognize them, and to react. And, of course, being a hockey goaltender was good training for learning the heartbreaks of exploring Mars.

So, I'd like to try to take you forward a number of years, and the exact date we can't say. But the time when people and machines, women and men and machines, are able to explore Mars. And, in this case, I think the question is really not one of *ifs*, but really one of when the time is right, when those benefits can be matched against the exploration risks. So I'm going to take you through some history to try to talk about that.

I think we all learn to explore in different ways. Shannon gave us a wonderful story of exploring in space. My beginnings in this regard came trying to look for Mars on Earth

James B. Garvin
NASA Chief Scientist for Mars and the Moon

In his position (since 2000) as chief scientist for NASA's Mars Exploration Program and the newly initiated lunar exploration initiative, James Garvin is responsible for formulation of scientific requirements for the exploration of Mars and the Moon and overall leadership of the scientific trajectory of the programs. He began his NASA career in 1984 in the Geodynamics Branch of NASA's Goddard Space Flight Center, Greenbelt, Maryland. He has served as chief scientist for the Shuttle Laser Altimeter experiment (SLA), which flew in Earth orbit twice, and as principal investigator on 12 airborne laser altimeter missions to such remote locales as Iceland and the Azores. He is a co-investigator on the Mars Orbital Laser Altimeter (MOLA) on board NASA's Mars Global Surveyor spacecraft and the NEAR Shoemaker laser rangefinder experiment. He has published over 60 peer-reviewed research articles and thousands of extended abstracts. Most recently (January 2004), he appeared on *The Late Show with David Letterman* as NASA's Mars expert.

through the eyes of the Viking mission that I was lucky enough to be an intern for. And in doing so, I recognized that there are a lot of difficult-to-measure things that, in fact, pose risks to the science and understanding we're trying to build. Earth as a training ground was one of the ways in my educational experience that I tried to accommodate risk. But I realized in doing so that there were things I couldn't measure. I didn't have the technology or the tools. Just as many of the explorers taught, I needed more tools to get more data. One way to obtain these data is to benefit from human spaceflight to make robotic measurements.

I realized that the landscapes that we want to understand on Mars and on Earth needed to be measured with new scales, robotic scales rather than human. I was fortunate to benefit from human spaceflight carrying my experiment, the Shuttle Laser Altimeter (SLA), for the first time in the mid-1990s, to measure part of the Earth. This gave me the benefit of human exploration, the Shuttle program, carrying a robotic instrument. The first landfall that instrument made, so to speak, was, rather ironically the path right over Mauna Kea, the truly largest mountain on planet Earth considering the depths from which it rises from the ocean floor. We were rather shocked, in fact, to realize that our flight path allowed us to measure this 4,000-meter place that I had visited as a graduate student some ten years before to learn how to measure Mars.

Now, in viewing human spaceflight to enable robotic experimentation, the same sort of reducing risk routine is important. The advantage of human spaceflight was that it was more akin to flying an aircraft experiment than many of us in the remote sensory arena are experienced with. We had people on site to fix the problems. On our particular flight, in one case, the switch that enabled the high-energy laser to make the measurements was, in fact, wired incorrectly. Human error, part of risk. So we had to command our crews, in this case, to flip the switch off to turn the laser on, and we had a very simple procedure, one step. We were extremely good at it and had we not had that capability, we would not have been able to fly the experiment.

Now, what did we learn? Well, we learned, as we did on Mars, that one of the big side benefits of exploration in the face of risks is the serendipity of discovery from making new measurements. Yes, we measured the shape of the Earth at scales of a few feet from the Shuttle orbit. We told the Shuttle command we were actually giving them orbits within a few feet during flight, which was quite compelling to them. But, we also managed to measure the heights of the vegetation, part of the dynamic carbon cycle of Earth. We had the capability to make measurements of Earth, which we're now making from the ISAT satellite as part of the Earth Observing System, that would help tell us about the carbon cycle on our own planet as we got ready to carry this kind of instrument to Mars to help prepare for landings of vehicles like the rovers.

So, my dream, as I became an explorer off the planet, was to map Mars in 3-D at the scales that humans and others would want to build. The science, of course, was to understand the history of the crust from the evolution of the planet, but also to bring Mars into closer focus, to allow us to make some of the views that allow us to imagine going to places that are very complicated and unique. Without the topographic perspective, learning how to fly into that kind of environment, with robots and then with humans, would have been impossible.

We've learned from the legacy of our forerunners, as we always do in exploration. I think it's important to go back to the Moon, as we are. But it's also important to go back in history to the Moon. Because the Surveyor Program, one of those antecedents built into Apollo, told us many things about the planet, and it was originally conceived as a risk mitigation step, not as a science mission. In the case of Surveyor, they were able to get to new places.

Surveyor 7 took the only picture from the rim of a gigantic fresh crater on the Moon—Tycho. It gave us a new vantage point on the Moon. It helped us look to places we might like to send human beings and understand that great world. These robotic forerunners were the steps that allowed us to get the first successful landings on another world, in this case, on Mars.

WILL THE RISKS BE ACCOMMODATED TO SEND HUMAN BEINGS? TODAY OUR ROVERS HAVE DRIVEN SPECTACULAR DISTANCES ON MARS AGAINST ALL ODDS. AND YET, IN 20 HOURS WE DROVE MANY TIMES FURTHER WITH A HUMAN SYSTEM. THOSE CHOICES ARE VERY IMPORTANT.

Viking collected 10,000 images from the surface, hundreds of spectra, 14 experiments of the most grandiose nature in the 1970s. Viking's first pictures were not beautiful vistas and landscapes; they were pictures of the feet of the vehicle. They were taken as part of a science contingency plan, the same kind of plan that we, in fact, asked Steve Squyres to implement in sending out his prize rover, Spirit, to Mars. And these pictures actually gave us science, the first views of the sort of fingernail scale or hand scale of Mars. And they showed the fact that Mars dust was kicked up and ended about an hour after landing. So this was science. Viking took science to new extremes in searching for the first chemical antecedents of life and doing other experiments. It also took great pictures. And exploration is about mapping ourselves into a new vantage point.

Now, some of these pictures were high risk. I remember vividly Jim Martin, the project manager, saying, "You're not taking any sunrise pictures. It's too cold. You're not going to do that." Well, we did. Scientists prevailed against all odds. We took that sunrise picture in an attempt to make Mars look the way our eyes would see it. Now, this was very controversial, but it's the way science works. We were trying to understand the Mars that we would see if we were there. That's part of exploration.

Mars has not always been easy. I could go on and on and talk about the graveyard of vehicles on Mars, from our great colleagues in the Soviet Union and our own Mars Observer, the great reconnaissance that we hung our program on in the 1980s. The failed Mars Polar Lander, a wonderful mission. The Beagle, that had the hopes of many to get to Mars. But out of these failures have come lessons—lessons that are the tough lessons of exploration; in this case, with robots before the people.

One of the lessons is that the polar regions on Mars are important. So we've selected a mission called Phoenix, after a year and a half long Olympic-class competition, to go back to recover that science. Likewise, our Mars Global Surveyor, from the ashes of the Mars Observer, has been monitoring Mars and mapping it, enabling landings of vehicles like our rovers. It's important to look at the legacy.

THE LADDER TO THE MOON WAS BUILT. WE WENT MULTIPLE TIMES,

LANDING SIX TIMES. THE SOVIETS WENT AND RETURNED WITH SAMPLES.

MAYBE THAT SAME LADDER IS NEEDED FOR MARS.

We built the system for Global Surveyor to operate for one Mars year. That would have been about 9,000 orbits. Today we are over 26,000 orbits. So we were able to mitigate the risk and continue the exploration—just as the rovers are.

So, I think it's important to understand that we mitigate risks by trying new vantage points, to know better, to be more informed. Today, we're building, constructing, innovating, and testing, at Lockheed Martin, the Mars Reconnaissance Orbiter—the ultimate reconnaissance step that will help pave the way for [the] future of Mars exploration. This is a complicated system; it's the largest reconnaissance orbiter to go to Mars in our history. We launch it next summer. It's the team of people that will help us mitigate those risks.

Now, for a minute, I would like to get off Mars. And, yeah, I may be a Martian here, but I think it's important to look at how exploration evolves. As we thought about the planet Venus in the late seventies, we had a vision for a complicated mission to map the planet and its atmosphere. Fifteen years later, that mission is realized in a different way. Thanks to technology, computer science, and information technology, the Magellan mission gave us higher resolution than we had imagined in the '70s, done in a different way. It allowed us to couple what we were seeing from the surface of this hellish world where our Soviet colleagues had landed 12 times [1969–81], to the big picture. This is sort of inverse exploration.

Now, the approach taken by our Soviet colleagues for their brief foray to the surface of the planet Venus, 450 degrees centigrade, was to use an approach

168

that was overdesigned to handle any environment. Overdesign the system to mitigate the risks of the unknown there. Surprisingly, they have not been back since their tremendous successes, which culminated with the Vega landings in 1986. But Venus still offers us a lesson in exploration.

I think the lessons from Mars are several. One is from our own proving ground here on Earth. Here and now, NASA is investing in programs to use Earth analogies, chemical process, laboratory scale analogies, to do science. Likewise with the Moon. The legacy of Harrison Schmitt and Jim Lovell and all those guys that went to the Moon and got those data, is that it is important to move ahead with the Moon as we start to learn about Mars.

Now, I thought it would be instructive to talk for a minute about the differences between the cooperative robotic and human exploration. It's important to remind you of a few facts. Facts are always good when we look at risk. One fact is what we did with the Apollo missions, which I would maintain were at least science enabling, however you choose to look at that. What we did was, we were able to interrogate the surface of another planet, even on foot with minimal tools, in an extremely short period of time. In two days, eight hours of being out on the lunar environment, we traveled two and a half kilometers. That's a human scale of interrogation of another planet. We touched and collected 50 pounds of rocks on the Apollo 12 mission to the Surveyor site. On Apollo 17, we upped the ante. In 20 hours of EVA we drove 36 kilometers.

Today, the pace of exploration is different. It's not different in its yield; it's different in its pace. The question of the timing is one of the important ones as we look at when will the time be right. Will the risks be accommodated to send human beings? Today our rovers have driven spectacular distances on Mars against all odds. And yet, in 20 hours we drove many times further with a human system. Those choices are very important.

Today on Mars we've experienced many things. We've looked at small craters in new ways, with robotic assets. We've driven in and are living inside one and yet, we hunger for more. Some craters at the scale of large football fields on Mars actually present tremendous exploration risk challenges. So big that during the Apollo era, the flight rules did not allow the crew to venture into the fresh impact craters of the Moon. Yet today, we have roving capabilities on Mars that could enable that. So it's that cooperative robotic and human exploration that's so important. In my own case, I think visiting impact sites has helped train me intellectually to understand some of those on Mars.

So, as we look at places to go, as we focus our attention on where the people and the machines need to go on Mars, we also need to learn from our experiences. One of the lessons of exploration is the risk of not exploring. Here on planet Earth we have a template for understanding the record of cosmic collisions, but the Moon and Mars offer a better template. It's the template of our history, and yet the opportunities for learning come both here and there. This is one of the learning factors in exploration. The risk of getting there—the trip, going to the places where the action is—can be mitigated by learning about places on Earth that can train us. Training is important. We've heard that again and again. Training with robots here on Earth, people, and then both on Mars.

Finally, I want to relate a story that I think is part of what makes science and exploration exciting. Some of the things we are going for, whether they be supernovae or understanding Mars or aspects of our Earth, are ephemeral. They will be gone. The atmosphere of Pluto is an example we talk about often in science.

I have been fortunate enough to visit a small volcanic island named Surtsey, born 40 years ago. It is already 25 percent gone. It may not survive this coming century, and, yet, it is a little microcosm of how the Earth responds to all the dynamic forces that shape landscapes on Earth. But it's a training ground, too. On this little island operates one of the types of processes that may make the ubiquitous gullies and hillsides on Mars, and we can go visit the island in the same chemical environment as those rocks on Mars. The time-lapse photography has been sped up. Instead of at Mars scale, this is at Earth scale. We can go visit, and, in a period of years, we can watch it evolve, measure [it], and understand how to explore it. We can also learn from new vantage points.

At NASA, it's important to empower the community to competitively seek ways to see Mars in new ways. This last couple of years we had a competition

... IT'S REALLY IMPORTANT TO NOT WAIT TO WONDER. THAT'S WHAT EXPLORATION

IS ABOUT. DON'T WAIT, BECAUSE IT'S IN THE GOING THAT YOU HAVE TO GO."

for the first Mars scout. One of the missions proposed by Joel Levine and his team was to look at Mars from air, to get around more, to do the recon closer to the ground of the Martian system, including the trace gases, that would help us be better informed. Being better informed thanks to reconnaissance has always made a difference in exploration.

So I think there is a set of converging pathways. The timing of the convergence will tell us when the risk can be accommodated to put people on site on Mars for the good of science. There are many pieces, and you can see them. We're doing some of them now: reconnaissance, sample selection with our rovers, understanding the things we see on Earth. Just this summer, I should say austral summer, a field team collected a new Martian meteorite in the middle of the range of Antarctica: a piece of Mars sent by Mother Nature to inform us about what we need to learn about. This collective approach is a way to reduce risk, and, by having a program that does so, we can learn. Where are we going with humans? Well, I hope it's exploring at least, in part, in the name of science.

I will finish with two minor quick thoughts. One is that sometimes exploring is better captured in the eyes of the artists. Georgia O'Keefe, at the dawn of the space age, painted a great picture, *The Ladder to the Moon*, from her vantage point in New Mexico. I think it was Taos. It was kind of an interesting flight approach to getting to the Moon that only a modernist could do. But, I think it is the epitome of the inspiration that allowed us to actually achieve that vision. The ladder to the Moon was built. We went multiple times, landing six times. The Soviets went and returned with samples. Maybe that same ladder is needed for Mars.

So, as we have all said during this conference, it's inspiring and, in fact, more than inspiring, catalyzing the youth to tell us how to go that's important. I think perhaps all the vision we talk about is a powerful risk mitigation tool.

I will leave you with one last thought as best I can. I was giving a commencement address to Thomas Jefferson High School in Virginia this last June. The students were really empowered. They wanted to do space exploration. They cheered when they saw a NASA person show up. I was stunned. I thought there was a rock star somewhere, and I couldn't imagine they were cheering for NASA. I thought, Wow! Here are five hundred of our best and brightest boys and girls wanting to do this. I stared at them, and, rarely for me, I was brought to a lack of words. I stopped a moment, then said, "You know, it's really important to not wait to wonder. That's what exploration is about. Don't wait, because it's in the going that you have to go." So we mitigate the risks by going and not waiting intelligently, and that's what we're doing now in our Mars program.

Back to Contents page

Mars Exploration Rovers

Exploration Rover (MER) mission, the mission of Spirit and Opportunity, and the risks that we took with that mission. I think by any standard, MER has to be looked at as one of the riskiest and one of the most complex robotic missions that NASA has ever undertaken, but it has been successful. We talked yesterday about mountaineering. Well, Spirit is now the first Martian robotic mountaineer, ascending the Columbia Hills. We talked this morning about oceans, and Opportunity is now exploring the remains of an ancient salty sea on Mars. Penny, I'm sorry, we haven't found any caves yet. Caves are kind of scary places if you're a solar-powered rover. So, we're probably going to stay away from those.

I'm going to talk about the risks that we took to make that success happen. There is one point I have to make from the very outset. It is so obvious that I almost don't need to say it, but it's also so fundamental that I have to say it. That is, there is a very, very fundamental difference between our mission and most new missions we are talking about here. When our rockets lifted off from Cape Canaveral last summer, our lives were not on the line. Now, there were a few meetings at NASA Headquarters where I wasn't quite so sure

Steve Squyres
Professor of Astronomy, Cornell University and
Scientific Principal Investigator for the Mars Exploration Rover Mission

Steve Squyres is the scientific principal investigator for NASA's Mars Exploration Rover Project. He received his Ph.D. from Cornell University in 1981 and spent five years working for NASA at their Ames Research Center before returning to Cornell, where he is now a professor of astronomy. Scientific research for which he is best known includes study of water on Mars and of a possible ocean on Jupiter's moon Europa. He has participated in many missions of planetary exploration, including the Voyager mission to Jupiter and Saturn, the Magellan mission to Venus, the Mars Odyssey, Mars Express, and MRO missions to Mars, the NEAR mission to the asteroid Eros, and the Cassini mission to Saturn. Squyres has served as the Chairman of NASA's Space Science Advisory Committee and as a member of the NASA Advisory Council.

about that, preceding launch! I almost feel like I don't belong up here with people like Shannon and with [Harrison] Jack Schmitt, but I think that our experiences do have much to say about how one takes risks in spaceflight, including human spaceflight. There were many aspects of our mission that are in common with what goes on in human spaceflight. We had a very challenging schedule. We had a very daunting technical task. We had an enormously large and complicated team to pull it all off. Addressing and aggressively mitigating the risks that come with all of those things is something on which we spent an enormous amount of time, and I think some of our lessons there do carry over to the very demanding realm of human spaceflight.

As Jim alluded to, our mission arose out of catastrophe. In 1998, NASA launched two missions to Mars. The Mars Polar Lander began its entry and descent sequence and was never heard from again. The most likely cause was determined to have been a single line of code that was missing that resulted, ultimately, probably, in the vehicle shutting off its motors about 40 meters above the ground and hitting the surface at about 50 miles per hour. Then, in reverse, the Mars Climate Orbiter was lost when a mix-up over English and metric units resulted in flying the spacecraft into the atmosphere and burning it up.

So, we were put in a position, which we all embraced from the start, of being involved in a mission that had to succeed. The credibility of a substantial portion of the Nation's space program and some of the institutions involved was very much riding on our success or on our failure. We had to come up with ways to address that risk that were commensurate with the expectations that had been forced upon us by circumstances.

As with any program, we addressed, and had to face, a wide variety of different kinds of risks. There was cost risk. There was programmatic risk. There were technology risks and environmental risks. There were operational and scheduling risks. I am going to address each of these briefly in turn. There were many things we did individually to mitigate each of those risks, but I think, almost above all, there was one thing we did from the start that addressed every single one of those risks. I alluded to this briefly in some remarks that I made yesterday. We knew what we were trying to do. We had a set of level-one requirements. They were negotiated with NASA Headquarters. They fit on a single piece of paper—two sides. They stated succinctly and clearly what the MER mission was expected to do. From the day that NASA said "go" to the day that we had a date on Pad 17A at Cape Canaveral was 34 months. We would not have made it had we not all had a clear, unambiguous, common understanding of what it was we were trying to accomplish. Those level-one requirements were our guide star.

I lost a lot of sleep wondering whether or not we were going to make it, but I never once questioned what it was we were trying to accomplish. We never had an ounce of uncertainty in our minds. That was tremendously enabling, because every time we faced a decision we turned to those requirements Do we do this test? Do we not? Do we include this component? Do we not? Does it help us meet the level-one requirements? If so, yes. If no, it's expendable. And it was that simple. I don't care how big or how small the organization, how complex or how

SO, WE WERE PUT IN A POSITION, WHICH WE ALL EMBRACED FROM

THE START, OF BEING INVOLVED IN A MISSION THAT HAD TO SUCCEED.

simple your task, I cannot overstate the importance of clear, unambiguous goal setting. It gives a crystalline clarity of purpose to your organization from top to bottom if everybody knows, with no ambiguity, what you are trying to achieve. That was fundamental to our success.

I am going to go through those risks that I listed.

Cost risks: When you get right down to it, our fundamental approach to cost risk was that, when we needed more money, NASA gave it to us. We originally costed the mission out at $688 million. We overran that by more than 100 million bucks. The reasons for those overruns are interesting, and I will be glad to tell any of you about them. They fundamentally had to do with some in-going assumptions that turned out to be flawed. Twice over the course of the development, Firouz Naderi, the program manager at JPL [Jet Propulsion Laboratory], and Pete Theisinger, our very able project manager, and I had to get on a plane and go back east and tell them we needed 50 million dollars more. The first time we did it, we were flogged. We then got our 50 million, and we promised never to come back again.

How long was it, Jim? About six months later, we were on your doorstep again. We were really flogged on that one. When it came right down to it, with so much on the line and so much at stake, the Agency was able to look at their priorities and say, "We have to make this work." Never once over the entire course of the MER development did we not do something important, something that was enabling of meeting our level-one requirements, because we didn't have the money. It never happened, and that was because the Agency made the commitment to make sure it never happened.

Programmatic risk: Programmatic risk means a lot of different things to a lot of people. I will define it rather narrowly to mean the way in which you interact with other programs over issues like personnel, facilities, and so forth. Our approach there, to be honest, was very much like our approach to cost risk. What we needed, we got. The Jet Propulsion Laboratory is an immensely talented, immensely capable organization, but their resources are not infinite. Whenever it came down to something critical—if we needed the right people, we got them. If we needed certain facilities, we got them. There just weren't any questions asked.

The team that was put together under Pete's leadership at JPL was the best that the Jet Propulsion Laboratory had to offer. Lab management always gave us everything that we needed. You can't do that for every project, obviously, but it is a matter of having your priorities straight. Your priorities were that MER had what MER needed, and what MER needed, MER got. There was a phrase around JPL that I heard about. Somebody would say, "I got MER'd". That meant that their facility or their engineer or somebody had been stolen away by MER to go off and make sure we got to Mars okay.

Technology risk: Our approach to technology risk is, basically, don't take any. Our mission was assembled almost entirely from existing, tested, proven technology. Air bags have been used on Mars, parachutes have been used on Mars, aerogel had been used on Mars. The payload was ready to go. The entire mission was put together from existing, qualified, capable hardware. Our computer was a smoking hot machine in 1985, okay, but it was good enough to meet the job that was laid out in the level-one requirements, and so that was what we used. You can sometimes accomplish extraordinarily innovative things by taking all the existing technologies and combining them together in novel ways. And I think there may be a lesson there; I don't think MER is the only opportunity out there for taking existing, proven, safe technologies and combining them together in ways that haven't quite been attempted before.

Environmental risks: This is a big one. There were many environments over the course of our flight over which we had little or no control and for which we had to do our best to prepare ourselves. Launch was an environment that was, as a spacecraft team, outside of our control. That was risky. Landing was certainly risky. Unless you have a fully deterministic landing system when you land on Mars, I don't care how much testing you do—you cannot build a perfectly safe Mars lander. You can build the best system you can, but you can always have one sharp pointy rock or one gust of wind that does you in, if you got unlucky that day at the landing site.

And so our approach to environmental risk was absolutely the best one that you can take: we built two of everything. Two rockets, two landers, two rovers, two payloads, identical up and down the line, but we built two of everything. This is a risk mitigation technique that does not carry over, obviously, into the realm of human spaceflight—you can't say, well, let's send two crews and maybe one of them will survive.

But if you have a robotic mission that must succeed, if you don't send two, you're crazy, in my personal opinion. It worked very well for us. And I'll also point out that it worked very well for the people who were involved in Mariner 3 and 4—Mariner 4 being the first successful Mars flyby, Mariner 3 going in the drink. The same is true for Mariner 8 and 9—the first successful Mars orbiter, when Mariner 8 went in the drink.

There's another aspect of environmental risk, which I think was not adequately appreciated by most people, and that had to do with risk to the science. We were going into a fundamentally unknown scientific environment. We did the best we could to select good landing sites, but we didn't really know what to expect. And one of my greatest fears when we actually first proposed MER to NASA as a single rover mission was that we would choose badly, that Mars would fake us out, and we'd get down on the surface and the science that we were seeking just simply wouldn't be there. If you had two rovers, and if you had a very diverse planet, as Mars is, you could send them to two very different sites, and maybe one of them is going to turn out to be the miracle site.

Mars did fake us out, by the way. If you had told me ahead of time, "Steve, one of the rovers is going to land on volcanic rocks and one's going to land on sedimentary rocks," and you'd said, " Gusev, Meridani" I would have said, "Yeah, sure. It's got to be volcanic rocks at Meridani and sedimentary rock at Gusev." It was the other way around. Mars completely faked us out. And the beauty of having that redundancy to mitigate that science risk is that if it really pays off and both vehicles get on the surface, you take advantage of that diversity to essentially double your science return, because you're in two completely different environments.

Operational risk: The chance that, when you try to do it, it's not going to work. There's no magic formula here, this one's really straightforward. You do it with margin and testing. Now, it's just down to block and tackling on this one. You build a lot of margin into your design and then you test and you test and you test and you test. And like I always say, you test it like you're going to fly it, and you fly it just the way you tested it. And we did a hell of a lot of testing on MER. Our schedule was all about testing. Everything that we did was about testing.

And in the end, those operational risks that we personally took paid off, and the margin in particular was very important. We put a lot of margin in the design—there's margin tucked away in so many nooks and crannies in that design you can't believe it. And it was that margin that made us comfortable signing up to a set of level-one requirements that says this vehicle will last for 90 Martian days on the surface. But if you've got that much margin in your pocket and a few things break your way, you might still be driving around on SOL 265, which is, I think, what today is. So margin pays off in big ways.

Finally, *schedule risk*: This was the worst risk that we faced, by far. In a very real sense, the entire story of the development of the Mars Exploration

Rover program—the development of Spirit and Opportunity—is the story of an extraordinary group of people facing schedule risk. Like I said, NASA said, "Go, and you've got to be there on the pad in 34 months." That was not enough time. It was not enough time. There were many things that we did to mitigate schedule risk—I cannot discuss them all. I will only mention two of them.

One of them—this will sound paradoxical, but it is not, and if you take anything away from what I have to say today, please get this point. Our schedule risk was mitigated to a great extent by the fact that we were flying two vehicles. That doesn't sound like it makes sense. It should be easier to build one than to build two. Well, under certain circumstances, if you're starved for people, starved for facilities, starved for money, then, yeah, that's true. But if you've got the people, if you've got the facilities, and you've got the money, then it helps to be hardware rich. You have more pieces on your chessboard and it puts you in a stronger position. Just as one trivially simple example, there are many tests that you run on vehicles like this that only have to be run on one of your two vehicles. And if you've got the facilities and you've got the people, you run those two tests—not in series, but in parallel, and you take up schedule. And we did that again and again and again and again.

And Matt Wallace, who was the manager of our ATLO—assembly test and launch operations—was a master. He was a hero of this mission and he played that game with those chess pieces with such intricacy and such skill that we made it, and I don't think we would have made it to the pad if he'd had only one vehicle. I think we had to do it with two.

The other way in which, I am somewhat ashamed to say, that we mitigated schedule risks is that we pushed an extraordinary group of people too hard. We pushed them beyond reasonable limits. It damaged people's health. It damaged people's relationships with their loved ones. We got away with it because we had an extraordinary group of people under an extraordinary group of circumstances, but that is not a sustainable approach to Mars exploration. You cannot go back to that well again and again. I do not believe that 34, 36, 38 months is enough time to do a robotic mission of that kind of complexity. I think you need 48, and I hope that lesson is one that is taken away from the MER mission.

I'd like to finish this on a slightly lighter note by telling you a story. We had a lot of discussion yesterday about humans versus robots. And as the robot guy here, I want to tell a story about the experience that I had that really taught me a lot about that particular topic. We were at first trying to figure out how to use a set of rovers on Mars to really do scientific exploration. The technology folks at JPL [Jet Propulsion Laboratory] built a wonderful little vehicle called FIDO. And FIDO was a great test rover—you could take it out in the field and you didn't worry about getting a few scratches in the paint.

We took it out to a place called Silver Lake in the Mojave Desert about 1997. And we went out there and it was the first time I had ever been out in the field. So I went out there with my team—a bunch of really high-priced geologic talent—some serious field geologists. And we got the rover out there and, of course, the rover breaks down. First time I've ever been out in the field, it's dusty, it's dirty, you know, the rover's not working. So okay, what am I going to do with all these bored geologists I've got on my hands? So I said, "Look, let's go on a geology walk. Let's go on a little field trip." So everybody got their boots and their rock hammers and their hand lenses and everything. And I picked up a notebook and a stopwatch. And we walked out to a nearby ridge where I knew there was some interesting geology exposed and we sat down—or rather I sat down—and they went off and they started geologizing.

. . . WHEN I HEAR PEOPLE POINT TO SPIRIT AND OPPORTUNITY AND SAY THAT THESE ARE EXAMPLES OF WHY WE DON'T NEED TO SEND HUMANS TO MARS, I GET VERY UPSET. BECAUSE THAT'S NOT EVEN THE RIGHT DISCUSSION TO BE HAVING. WE MUST SEND HUMANS TO MARS. WE CAN'T DO IT SOON ENOUGH FOR ME.

And I started timing them. You know, how long does it take for Andy Knoll to walk over to that rock? How long does it take Ray Albertson to pick that thing up and break it open with his rock hammer and look at it with a hand lens? And they were doing a lot of things that our rovers couldn't do, but I focused on the things they were doing that our rovers could do. And, you know, I did it as quantitatively as I could—this was hardly a controlled experiment. And when I looked at the numbers afterwards, what I found was that what our magnificent robotic vehicles can do in an entire day on Mars, these guys could do in about 30–45 seconds.

We are very far away from being able to build robots—I'm not going to see it in my lifetime—that have anything like the capabilities that humans will have to explore, let alone to inspire. And when I hear people point to Spirit and Opportunity and say that these are examples of why we don't need to send humans to Mars, I get very upset. Because that's not even the right discussion to be having. We must send humans to Mars. We can't do it soon enough for me. You know, I'm a robot guy. I mean, I love Spirit and Opportunity—and I use a word like "love" very advisedly when talking about a hunk of metal.

But I love those machines. I miss them. I do. But they will never, ever have the capabilities that humans will have and I sure hope you send people soon.

COBE and the James Webb Space Telescope

I want to talk to you about the Cosmic Background Explorer (COBE) satellite and the James Webb Space Telescope (JWST), to give you examples of two extraordinarily risky visions that I have worked on. One of them hasn't been launched yet and one was launched some time ago. So the concentration is on the James Webb Telescope, which used to be called the Next Generation Space Telescope. We had a lot of Trekkies at Headquarters and they were very proud to name it the Next Generation Space Telescope. It was renamed after NASA Administrator James Webb. I didn't know much about James Webb until I read a biography of him and he was in fact a remarkable person and it's a tremendous honor for the telescope to have his name attached to it. If you want to know more about it, there's a book called *Powering Apollo*. It is really very inspiring to read and also points out that he was really very interested in reducing risk by adopting and learning about new methods of management.

And management, I think, is our biggest risk in many areas. Many people have spoken before about losing concentration on the risks that we face and panicking in the dive or whatever it might be. We have to sort of keep the same focus all the way up to top of the management chain, otherwise we get in trouble.

John Mather
James Webb Space Telescope Senior Project Scientist, NASA Goddard Space Flight Center

John C. Mather is a senior astrophysicist in the Infrared Astrophysics Branch at NASA/Goddard Space Flight Center. His research centers on infrared astronomy and cosmology. As an National Research Council postdoctoral fellow at the Goddard Institute for Space Studies (New York City), he led the proposal efforts for the Cosmic Background Explorer (1974–76), and came to Goddard Space Flight Center to be the study scientist (1976–88), project scientist (1988–present), and also the principal investigator for the Far Infrared Absolute Spectrophotometer (FIRAS) on COBE. He showed that the cosmic microwave background radiation has a blackbody spectrum within 50 parts per million. As study scientist (1995–99) and project scientist (1999–present) for the James Webb Space Telescope, he coleads the science team with Peter Stockman (STScI) and represents scientific interests within the project management.

So, what do I think risks are? There are a lot of things that people call risks that aren't, to my way of thinking. Some things are intrinsically chancy, you couldn't possibly predict whether they would or would not happen. And we have done a lot of things to reduce those risks by working harder and harder on what you can control and predict. But, also, there are a lot of things that depend on who's working on it, who's thinking about it, who's paying. And if you're the management paying to reduce risks and you have another person breathing down on you saying, "If you spend any more money on that project, you're out," there are a lot of kinds of risk that people feel and take.

So why should we take risks? Well, exploration and science are always about the unknown, and that's intrinsically the nature of it. So that's why we're here.

The first project that I did was a mission to measure the primordial cosmic microwave background radiation in a couple of different ways, and to look for the accumulated light of the first galaxies. And we did actually succeed in all of these objectives with the COBE satellite. This project was remarkably risky, considering when it was proposed and what it was like at the time. This was proposed in 1974. I organized a team six months out of graduate school to propose this mission, and when NASA decided to take us on as a serious study, they were taking a risk on us. But we did actually get associated with a truly wonderful engineering team at Goddard Space Flight Center and they produced this whole thing. It was an in-house project, which is not one of the more common ways that we do projects at NASA. But it was a wonderful thing for this project.

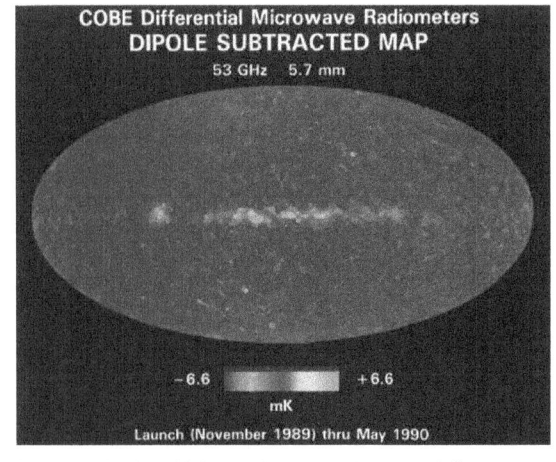

In 1974 people did not have computers on every desk. We did our first engineering drawings with pencils. And our calculators, our computers, were HP35s, and it was a miracle to have one. So people thought differently about risk because there was just too much you could never figure out or calculate. You did not have a finite-element model of everything you wanted. You just said, "Well, I think that'll probably work." And sometimes you were right and sometimes you weren't.

So then there was testing. This went through two metamorphoses. COBE was proposed first as payload for a Delta rocket. Then NASA went and put all of its eggs in one basket with a lot of kick from Congress and said, "We're killing all the expendable launch vehicles. We're going to send everything up on Shuttles." So this was redesigned to go on a Shuttle and it was a 10,000-pound spacecraft that had to go up from California. As far as I know, it was the only scientific payload that was to have the Shuttle launch from California.

Then the *Challenger* happened when we had more or less completed our design and were putting our spacecraft together. And it became pretty clear that there was not going to be a Shuttle launch from California ever again, and it wasn't just because the Shuttles were dangerous—there wasn't enough traffic for the purpose.

So we had to rebuild it all. We had also about a 30-month schedule we had to meet to rebuild everything. Fortunately, we did have the instruments that were more or less complete. And the business end did not really have to change much. But everything below that did have to change, and it was all new mechanical and thermal structures.

So our Deputy Project Manager took a risk. He started hunting around in the rest of the world for a foreign launch vehicle, and Headquarters informed him that he would lose some body parts if he kept on doing that, and that was a risk. Headquarters did, however, recognize that potentially this mission could be the first new science mission to go up after the *Challenger*. And so they said, "Well, if we could get you a Delta, could you fit?" So the answer was: "Yeah, just barely." And so the Delta was found. It had to be brought together from the spare parts left around in hangars elsewhere and pigeon droppings had eaten holes through the tanks in a few places. They were welded closed. So some of the stuff was a little bit of a risk.

But that was not the hard part. There were quite a few other kinds of risks in here. Of the business end, the scientific instruments, two of them were located inside a helium cryostat, and we had almost no experience with operating anything at very low temperatures. Certainly, our space engineering team did not know much about it. We all learned a lot going through this project.

. . . MY FIRST THING IS, IF IT'S NOT TESTED, IT WILL FAIL, AND

THAT YOU PROBABLY WON'T BE ABLE TO FIX IT EITHER.

Another really tough challenge that afflicted us seriously in the early days was that our budget was always limited. We didn't enjoy the virtue of being a top-priority project. So there were quite a lot of things that we did that were probably wrong, but we just knew we couldn't get the money to do the thing right.

However, that did change after the *Challenger*. And, so, the management approach changed. Charlie Pellerin came around and said, "If there's anybody on this project who knows any reason why this isn't right, tell us now." And he was pretty serious that we would have to tell him. He wanted to know the bad news, if there was any, because he needed to make sure this was going to turn out. So, I think this was an example that management attitude and our ability to raise funds to do the right thing were critically important to success.

Anyway, the whole story as best I could tell it is recorded in a book called *The Very First Light*. It's about ten years ago now, and it was written for a general audience. People told me afterwards they were out of breath from reading it because there were so many hazards that we faced and recovered from; and I think that's not unusual in the space business.

But it was worth it. We showed the spectrum of the universe is a perfect blackbody spectrum for a temperature of 2.725 plus or minus .001 Kelvin, which in its earlier incarnation brought us a standing ovation when we showed it to the American Astronomical Society. The question of whether the Big Bang theory was correct at all was still a somewhat open issue when this was reported. And to produce such a perfect measurement was tremendous cause for a celebration. And all the critics finally had to give up and agree that, well, maybe the Big Bang was really right. A map of the brightness of that background radiation over the entire sky was used to not only confirm the Big Bang story, but also to start in on the question of what's the rest of the universe like? The map and the details tell us basically that it's true—we're only 4 percent of the total universe, the matter that we know about. There's something like 20 percent more dark matter that still has attractive gravity. And we weren't quite sure then, but we were beginning to get onto it, that there's also a repulsion force that causes the acceleration of the universe. The universe is going faster and faster, and danged if I know what that's about! A lot of people have guesses, none of which we can confirm as yet. So this is a very open subject.

So these two results basically started up immense industries. There has already been a successor spacecraft for the measurement of the map and it's done far better. There's another one planned, and another one is hoped for after that. So we have made tremendous results out of this project, which seemed extremely risky technically.

Now, I want to go on to something that's perhaps easier in a sense, but much more difficult in another, the James Webb Space Telescope. This mission is the scientific successor to the Hubble Space Telescope, and it was conceived a long time ago, back in 1995, as the successor. And it's a scientific successor, but not a replacement. A lot of people are very concerned about what the future of Hubble is, and for anyone who cares, this is not a replacement.

This is not doing what Hubble does. This is looking farther away into the very distant universe and looking deeper into the places where stars and planets are being made, and it's much more challenging in a different way. To accomplish the objectives we need a much bigger telescope than Hubble. Well, how are you going to do that? Well, I'll show you. It also needs to be very, very cold because we need to see infrared light. The most distant universe is red-shifted, as it appears to us. The ultraviolet light that was emitted by those most distant things we want to see comes out in the infrared. And the visible light and other things come out in much longer infrared wavelengths. So we have to have a full telescope, and we are driven to a solution that happens in deep space. So we couldn't find any way around it. We sure knew that it would be great if we could service this mission as the Hubble had been serviced. We could not find a design that would allow that.

So we did negotiate a deal with the European and Canadian Space Agency and we did get the blessing of the National Academy of Sciences that this was the next big thing. It is a vision which strikes fear and trembling into the hearts of engineers, because the payload escapes from the Earth. It's going to go a million miles away from the Earth and when it gets partway out there it's going to deploy. And deployment in mechanical devices also terrified many engineers. We've had a lot of trouble with mechanical things and they're plenty right to be worried about it. It's got its solar arrays to deploy, then a solar baffle, a heat shield, then graphite fiber poles, a little bit stouter than fishing poles but not a whole lot. The shield pops open and becomes five layers. The five-layer shield is a much better thermal shield. Then a support tower is erected and the secondary mirror comes out and deploys on its linkages. The gold-coated primary mirror there is made out of beryllium hexagons, but it's coated with gold. And it deploys one wing of the hexagon, and now the other wing.

It's a million miles away. It's a million miles in the opposite direction from the Sun. There's a point called the Lagrange point L2 where it's a semi-stable orbit. And if you hovered around that spot, you could stay there with only a small nudge and move around the Sun with the Earth all year long.

So that's where we're going. This is the thing that's supposed to go there. And if you had asked us ten years ago if this was going to work, people would laugh at you and say, "Nah, you couldn't do anything like that." The company that's building this for us is Northrop Grumman. It's next to L.A. airport. And they tell us that they've actually deployed many, many things in space for other government agencies which they can't tell us about! But there is a reason why this technology was much more mature than astronomers ever imagined.

So, anyway, there's an awful lot of engineering risk in that thing. We had altogether twelve contracts to learn how to build the ultra-lightweight mirror that we need. The mirror is a chunk of beryllium, which is polished to the required accuracy. And we've proven that it will stay the right shape when it cools down. So this is a truly remarkable accomplishment. And, jeepers, it looks just like a mirror, doesn't it? But it took years and years of cooperative technology and competition.

Now, I want to show you the result as a formal tool that we use for analyzing risk. If we didn't have a formal tool I'm convinced we could never get there, because the way that we did risk analysis for the COBE project was: "Well, I think it'll work or I think it won't." Now we have a very formal process. We have a giant risk database. Our risk manager is here if you want to ask her more questions about it.

But we have engineers that fill out forms and we have weekly meetings, and we keep track of every single thing that we're worried about. And sometimes we retire it and sometimes we say, "Oh, it's getting worse," and we have to do something more vigorous about it. At least three of the top issues for us were questions about people—you know, can the agencies agree on something? So this is an example that's very typical, that some of the hardest problems are

185

negotiations. And I just wanted to emphasize that we have a method for doing this, and never to forget what the problems are.

I wanted to share some of my observations of physical things with you. And this, I have to admit, started with my learning experiences in school. I had a thesis project which failed on its first flight, and I learned a lot from that failure. And I know how we got into that mind-set that said, "Well, let's fly it anyway." We were tired, and we didn't have any more money. However, this thing did not work for three different reasons and so I learned something.

So, my first thing is, if it's not tested, it will fail, and that you probably won't be able to fix it either. And sometimes if it is tested, it will fail anyway, but at least you'll have a chance to fix it. But it will cost you. If you don't have a spare part or a backup plan, it will definitely fail. And if you only test it a little bit to see if it will do what it's supposed to do, then it'll do something else. So I've come to a similar conclusion to what Steve Squyres was recommending: You need lots of hardware around to work with, because things are going to go wrong and you need to be able to test out your idea on one thing while you're fixing the other one. So you need to have a lot of smart people thinking a lot about really terrible things, things that could go wrong and might just go wrong, and not being too limited to thinking about the things that you only know you can fix. Things that have the highest consequence will often be things that you missed because, you know, "Oh, I can't fix that." So you need lots of external review and we do have lots of external review. And so that's the number one thing.

There's another issue, about individual people. I don't think that human beings as a group are particularly good at balancing lots of likelihood and consequence. I know a lot of people [who] have fallen off of things and hurt themselves badly, including one of our senior managers on the COBE project— after he retired, fortunately for the project. So we're not really good at this. We need a formal tool. And we have a formal tool, but if you don't use it, you will definitely be in trouble! That's a conclusion from this.

So I think our greatest risk is lack of imagination. A lot of imaginary things you just have to explore. Once you've decided where you're going to go with what you get from your imagination, then you have to imagine all the things that could go wrong. You have to rehearse all that. If you were a performer, as my wife is, then you rehearse before you go out on stage. And people who are successful in our business rehearse and rehearse and rehearse, too. But I know that, at least in my history, we have been very easily blinded by thinking about what we have to work with rather than is it actually required? Nature doesn't really care whether we have enough resources to think about this problem. Either we did it right, or we didn't. If you didn't build it right and you think of this fact, then you better tell people and get the resources. Otherwise, you might as well not have started.

From the Earth to the Moon

I'm just a little-used screenwriter, and when I look at the people on this panel, I think I'm the answer to: "What's wrong with this picture?" The other thing that comes to mind is my *Mission to Mars* credit. There's a thing in Hollywood where you fight hard to get credit on something, because you'll usually get some money when the DVD sells. And you have to weigh the value of that money versus being humiliated in front of people at NASA for having been involved in a movie as bad as *Mission to Mars*. So—I really don't know if it was worth it. But anyway . . .

In 1996 I got a call from my agent and she said that Tom Hanks was doing a history of the Apollo program for HBO and did I want to be involved, and I said, "Sign me up." I read the outlines that they had prepared and I read Andy Chaikin's book. The episode that jumped out at me for dramatic purposes was the episode that, at that point, was then called "The Fire." It was later re-titled, for good reason, "Apollo I." They said, "Sorry, that's already taken by another writer." And the next day, I got a call that the writer had dropped out, so I got a chance to write that episode. In a very personally selfish way, that changed my career. Up until that point, I'd been an action writer. I did *Speed* and *Broken Arrow* and

Graham Yost
Writer/Director

Graham Yost is a writer/director from Toronto, Canada. His work includes the television series *From the Earth to the Moon* (1998), *Mission to Mars* (2000), and the 1994 box office hit *Speed*, starring Keanu Reeves.

those were fun movies, but this was the first time I got to write real people and really interesting and real dramatic situations. And I remember the highest compliment that I got was, at one point, Frank Darabont was going to direct the episode—Frank Darabont has directed *The Shawshank Redemption* and *The Green Mile*. He said while he was reading it, he kept on flipping back to the title page and saying, "This is the bus guy?"

As a little Canadian boy, I watched the Moon program from Canada and just loved it, which is why I said I wanted to be on board—but it made me a true space geek. The term on *Earth to the Moon* was "you've become a helmet-sniffer," if anyone knows the term from sports. That's why I'm here. And we would follow around Dave Scott, our astronaut adviser, and I remember telling my wife, "I just keep looking at his feet," because those feet were on the Moon.

In the writing of this episode, "Apollo 1," I decided very early on that I wanted to focus on Frank Borman, who was part of the Apollo 1—it was actually, technically, called the Apollo 204 Review Board. We're going to show a clip from the episode. So this is Frank Borman. It's later on in the episode and it's Frank Borman, played by David Andrews, who's testifying in front of a Senate committee. And I made Walter Mondale the bad guy, but that's a whole other story.

[Dialogue from video clip is indented.]

Senator Mondale: Colonel Borman, would you have entered the spacecraft on the morning of the accident if your turn had been called?

Frank Borman: Yes, sir.

Senator Mondale: Would you have had any hesitancy?

Frank Borman: No, sir.

Senator Mondale: Were there defects in workmanship?

Frank Borman: There were.

Senator Mondale: And did these defects go beyond workmanship?

Frank Borman: Yes, sir, there were defects in design.

Senator Mondale: If you had entered that spacecraft on that morning, would you have been motivated by a desire to take risks?

Frank Borman: No, sir. Sometimes there are romantic, silk-scarf notions attributed to this business, but we're professionals. We will accept it, certainly, but not undue risks.

Senator Mondale: Let me rephrase the question. Knowing what you know now, would you have entered that spacecraft?

Frank Borman: No, sir.

Senator Mondale: Colonel Borman, how did Commander Grissom and his crew feel about the readiness of the vehicle?

Frank Borman: I talked to Ed White shortly before the accident.

He thought they were over most of their problems and were on their way . . .

Senator Mondale: Didn't Commander Grissom once hang a lemon on the simulator?

Frank Borman: You had to know Gus.

Senator Mondale: Did Commander Grissom hang a lemon on the simulator?

Frank Borman: Yes, sir.

Second Senator: [interrupts Mondale] Tell us about him, Colonel. Sorry, Senator, I just have a couple of quick questions. Would you yield for a minute or two?

Senator Mondale: Actually, Mr. Chairman, I—

Second Senator: Thanks. Colonel Borman, you just said, "You had to know Gus." And I think that that's been missing in here the past few days. I'd like the record to contain just a little about the men who perished in that fire. Colonel, could you do that for us?

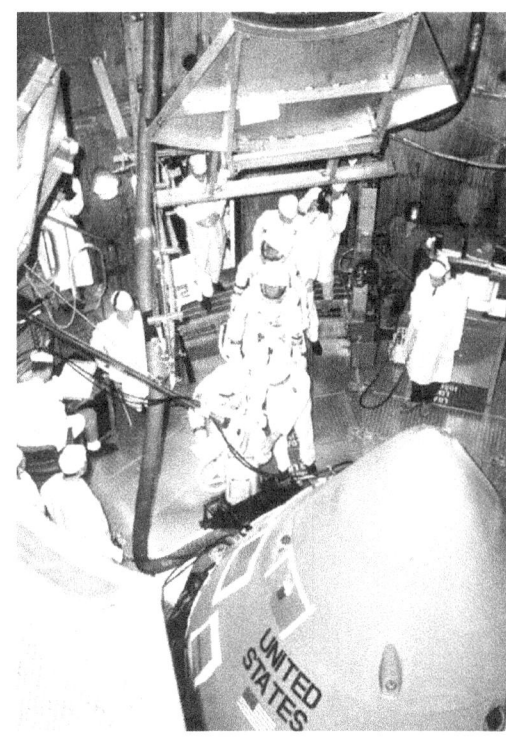

Frank Borman: Gus Grissom was the first astronaut to be asked to fly three times. Mercury, Gemini, and Apollo. He loved being an astronaut, except for the publicity and display that comes with the job. There are no front windows on the house he built for Betty in Timber Cove because he didn't want people looking in. If that gives you the impression that Gus was a cranky SOB, well, he was, at times. But I would have trusted him with my life.

Ed White was a big man for an astronaut, a shade under six feet. As you well know, Ed was the first American to walk in space. There's a story going around that when he was on his spacewalk, he stayed out after he had been ordered in because he was having such a good time. Funny story, but it would have meant Ed White disobeyed an order. Not going to happen. Ed was a West Point man. Duty, honor, country were not just words to him. He was one of my closest friends.

Roger Chaffee, I didn't know that well. He was one of the new guys, very energetic, very excited. I heard a story about him, though. He was out on Long Island visiting the Grumman facility where they were building the descent stage of the lunar module. He saw a group of men standing in the corner. He found out these were the fellows that make the tools that make the machines. None of the big wigs that were escorted through there ever talked to these guys. But Roger went over and made them feel like they were the most important part of the program.

Second Senator: Colonel, this isn't a court of law, so I can ask you something that's completely hypothetical. If you could somehow reach beyond the wall of death and talk to Grissom, White, and Chaffee, what do you imagine they would say about the fire?

Frank Borman: I was—I was hoping that someone would ask that. I don't know what Roger or Ed would say, but I can let Gus speak for himself. Back in January, he talked to a group of reporters. They asked him about the dangers involved in going to the Moon.

[Additional indentation to set off reminiscence of Grissom speaking.]

> *Gus Grissom:* We're in a risky business, and we hope if anything happens to us, it will not delay the program. The conquest of space is worth the risk of life. Our God-given curiosity will force us to go there ourselves, because in the final analysis, only man can fully evaluate the Moon in terms understandable to other men.

Second Senator: Colonel, at the risk of being gruesome, we've heard about the fire from everyone who was there, everyone except the astronauts themselves, of course. Can you tell me what they went through? What it was like for them?

Frank Borman: I can only tell you what we know or, at least, what we think we know. When it happened, they were just waiting for the test to resume.

[Additional indentation to set off flashbacks to astronauts caught in Apollo I fire.]

> *Gus Grissom:* How are we going to get to the Moon if we can't talk between three buildings? I can't hear a thing you're saying. Jesus Christ, I said, how are we going to get to the Moon if we can't talk between two or three buildings?

Frank Borman: They didn't see the spark that caused the fire because it was behind the panel door, down below Gus's feet. Because of the oxygen, the spark was able to jump out into the netting under the seats. Gus probably saw it first because it was closest to him.

> *Astronaut:* Fire! We have fire!

Frank Borman: Procedure would have had Gus push down Ed's headrest so that Ed could have started turning the latches.

Astronaut: We have a bad fire! [pause] Hurry up!

Frank Borman: Now, it just took me a minute or more to tell you all that. In actuality, from the first mention of the fire to the rupture of the hull only 15 seconds went by.

Second Senator: Colonel, what caused the fire? I'm not talking about wires and oxygen. It seems that some people think that NASA pressured North American to meet unrealistic and arbitrary deadlines and that in turn North American allowed safety to be compromised.

Frank Borman: I won't deny that there's been pressure to meet deadlines but safety has never been intentionally compromised.

Second Senator: Then what caused the fire?

Frank Borman: A failure of imagination. We've always known there was the possibility of a fire in a spacecraft. But the fear was always that it would happen in space when you were 180 miles from terra firma and the nearest fire station. That was the worry. No one ever imagined that it would happen on the ground. If anyone had thought of it, the test would have been classified as hazardous. But it wasn't. We just didn't think of it. Now whose fault is that? Well, it's North American's fault. It's NASA's fault. It's the fault of every person who ever worked on Apollo. It's my fault. I didn't think the test was hazardous. No one did. I wish to God we had.

Second Senator: Now before we all go home, is there any statement you personally would like to make?

Frank Borman: I think I'm safe in speaking for all the astronauts when I say that we are confident in our management. We're confident in our training, in our engineering, and in ourselves. The real question is, are you confident in us?

Second Senator: What do you think we should do Colonel?

Frank Borman: I think you should stop this witch-hunt and let us go to the Moon. [Applause]

Second Senator: Senator Mondale, back to you.

Senator Mondale: Thank you, Mr. Chairman. I have nothing further.

Second Senator: Thank you, Colonel.

[End of video clip]

John Grunsfeld: I'm going to exercise my executive privilege here for just a minute, and I'll let Graham continue, but, Graham, this is why we invited you. I can't help watching that without getting a little tear in my eye. I've been up at NASA Headquarters for a little over a year. When I came to Headquarters after the loss of *Columbia* it was because of the pain that I felt for the crew and my friends and this kind of circumstance. When I started watching this video

I thought, welcome to Mr. O'Keefe's world. And Bill Readdy's world. And all of us here from NASA who had to suffer through the loss of *Columbia* and crew. This really does address the central issue that we're here to discuss, which is how do we decide, when do we decide to go on, given the loss of our friends, for something we all believe is crucially important personally, professionally, for the planet, and for our friends.

I know Mr. O'Keefe has to leave in just a little bit, but I'd like to take this opportunity just to thank you as, hopefully, folks thank James Webb for the perseverance, the energy. For those of you who [don't] know me, I am an intense workaholic. I can look to a few members in the audience who are shaking their heads in agreement and Mr. O'Keefe, you're the first person that I have been unable to keep up with. Thank you very much. Thank you very much for helping sponsor this risk symposium.

I think in a nutshell that things like that video clip show the key to communicating this risk to the public effectively. So, back to you.

THIS REALLY DOES ADDRESS THE CENTRAL ISSUE THAT WE'RE HERE TO DISCUSS, WHICH IS HOW DO WE DECIDE, WHEN DO WE DECIDE TO GO ON, GIVEN THE LOSS OF OUR FRIENDS, FOR SOMETHING WE ALL BELIEVE IS CRUCIALLY IMPORTANT PERSONALLY, PROFESSIONALLY, FOR THE PLANET, AND FOR OUR FRIENDS.

Graham Yost: Thank you very much. After working on *Earth to the Moon*, I was looked at in Hollywood as the guy who, if NASA ever had a problem, would write the thing about it. So I became the disaster guy. This was also incredibly tough. I wrote a screenplay in '99 on *Challenger*. That has never been produced. Partly because, I think—it was for 20th Century Fox—and they were looking for white hats and black hats, and what I found was human beings.

Then, I got to work for HBO on a thing on *Mir* and looked at Jerry Linenger's experiences up there and the fire that they had. Also Mike Foale and the docking incident. At any rate, though, *Earth to the Moon* was the focus for me in risk. We shot it mostly in Florida. We did the lunar surface stuff outside of L.A., but we shot the rest of it in Florida. Our joke at the time was, just like the Moon program, we're thousands of miles from home, we're spending way too much money, and it's taking too long. The difference was, and it's been mentioned about the robotic missions, there were risks to career, risks to family, but there really, ultimately, was no risking of life in doing a miniseries for HBO and hanging out with Tom Hanks. That's not a hard thing.

But when I think about risk as I've heard over the past couple of days, I'm reminded of George Carlin's famous line—judging risk is very subjective.

George Carlin's line about driving was, "Have you ever noticed, anyone driving slower than you is an idiot and anyone who's driving faster is a maniac?" In hearing some of the things, we feel like we can somehow judge our own risk level. We know what we can handle. The classic subjective thing is flying versus driving. We know statistically flying is lot safer, and, yet, somehow, we feel that if we're in charge, we can handle that risk. I remember when I was living in New York and some crime had happened to a stranger. That was the thing we were always concerned with, stranger on stranger crime. You would find out when it happened and where it happened and you'd say, "Oh, I never would have been there." So, it's not risky.

In terms of Apollo, as I said, one of the great honors of doing *Earth to the Moon* was meeting the astronauts and spending a lot of time with Dave Scott. I got to direct the episode about Apollo 9. So I spent a lot of time with him talking about that. I also worked with him getting the script ready for the episode covering the Apollo 15 mission. He told me that there was a big discussion about what the rover walk-back limit on it would be, how far the rover could go before, if it broke down, they would have to walk back to the LEM. The proposal was that they should have—I forget the term—it was like a double walk-back limit or something. Because, what if the rover failed and one of the Portable Life Support System (PLSS) backpacks failed, that they should be able to go back on one PLSS backpack. Dave said, "No. That's just going to hamper us too much. That's going to hamstring us. We need to go as far as the single walk-back limit." I said, "Well, what would have happened if you had a PLSS failure and a rover failure?" He said, "Well, we would have had a bad day." That was his perspective, and that was his choice.

Thinking about risk and NASA and space exploration, you have to realize that people like Shannon Lucid and people like [Harrison] Jack Schmitt and the other astronauts, they're perfectly capable of judging whether or not something is safe. Just like David Andrews, Borman is saying: We know what we're willing to take. The reality is that space exploration, unlike the *Magellan* voyages, has been a volunteer thing. There's been no torturing of astronauts and telling them that they have to go into space. Which again, may not be a bad idea. [Laughter] It's important not to rule anything out.

This does bring up the other project I worked on in which that was always one of the questions. Judy Resnik and Dick Scobee and Elison Onizuka knew what they were dealing with in spaceflight but did Christa McAuliffe really know? Did Greg Jarvis really know? The thing is, they were told. They were told as well as anyone can communicate to them. It's not about statistics. Dick Scobee told Christa McAuliffe the classic line, "When you launch the Shuttle, everyone is at least three miles away except for us. We're going to be sitting on top." For me, in researching *Challenger*—and again it's probably one of the reasons it didn't get produced—was that the problem with *Challenger* wasn't that NASA somehow got lax with risk. There are all these theories, by the way. I don't know if any of you have read these books, but that NASA was pressured to make the launch in order to meet up with Reagan's State of the Union address that night and all this stuff. The future of space was in Reagan's hands, that there was pressure, pressure, pressure. I think that's absurd, when you look into it, and that's what Dave Scott would call an "outside the culture" view of it. When you get inside, you realize it was just people doing the job they had and the best job they could. Everyone working on the program knew the astronauts or met them at some time. There was no laxity in NASA on risk.

To me, in looking at it, the problem was with public perception of risk. This has come up again and again over the past couple of days, but because I'm last I get to say it again anyway. What can the public tolerate? What are they expecting? I think that when *Challenger* happened, NASA was a victim of its own success. If you consider Apollo 13 was a close call. We've heard that term mentioned, but other than that, it was just a string of successes with manned missions. The expectation in the public rose. There was also media pressure—the media fed into that. There was a classic tape of Peter Jennings on the 26th of January—the night before *Challenger*—saying another on-time departure is too much of a challenge for *Challenger*, because there was a socket wrench that they couldn't work.

But, the truth be told, NASA at that time was part of the problem, because NASA had promised that the Shuttle was going to be a routine access to space. As anyone who I've talked to involved with spaceflight knows, there's no such thing. It's not routine. It's not—as Mike Foale said last night—it's not flying a big aircraft. It's something far more complicated and far more risky than that. During the Presidential Commission on *Challenger* a figure came up and I don't know the source of it so, if it's not true, forgive me, but it has been said that the Shuttle stood only a 1 in 100,000 chance of having a disaster. It was Richard Feynman, who was on the commission, who worked out the simple math that that means the Shuttle would launch once a day for three hundred years before something happened. That was an unreal expectation, an unreal offer to the public, that it's going to be that good, that sure.

The thing is that we have public accountability. We have a transparent program. There are problems with that, but I still think the good outweighs the bad. In researching the NASA stuff, we also, at one point for *Earth to the Moon*, we were going to do a special two-hour episode about the Soviet program. It just became too expensive. One of the things that we found out is that there were horrific accidents. The testing of an N1 with over 1,000 people killed in one explosion. No one ever heard about it. No one in North America ever heard about it. No one in Russia heard about it.

In fact, I would say that culture of secrecy is something that, as Mike Foale said last night talking about the docking crash on *Mir*, contributed to that: The idea that we don't have to share everything. We don't have to tell you everything. It's all okay.

My closing thoughts have to do with the question of humans versus robots. A lot of people have said it's kind of an absurd question. They have to go hand in hand. To me, humans versus robots is, frankly, not about risk. Ultimately, it's about money. My feeling in having written about space exploration is that the notion of risk is almost secondary to whether or not we move forward into space. I think the public will bear whatever the risk is because they know that the people involved will bear it. Because the astronauts sign up, because it is voluntary. The question becomes, is the public going to get behind it? The public is important because they're paying for it.

Even though I'm Canadian, I pay American tax dollars. So, it becomes a public concern. If something happened back in the 1960s to Lockheed or Grumman or a North American test, well, that was private enterprise. That's okay. That's their deal. That's their pilot. But when it's our pilots, when it's our astronauts, then it becomes something that people have to get behind.

Dave Scott told me that Neil Armstrong once gave a presentation saying—and this is sort of a gloomy note to end on—but going to the Moon was really the convergence of several important things. The technology was available, the money, and there was the public will. And I think the big question is going to be

MY FEELING IN HAVING WRITTEN ABOUT SPACE EXPLORATION IS THAT THE NOTION

OF RISK IS ALMOST SECONDARY TO WHETHER OR NOT WE MOVE FORWARD INTO

SPACE. I THINK THE PUBLIC WILL BEAR WHATEVER THE RISK IS BECAUSE THEY

KNOW THAT THE PEOPLE INVOLVED WILL BEAR IT.

to get the public will to go back to the Moon and on to Mars. And I don't know if it's just a matter of communicating it. I think it is also a matter of, somehow, in the zeitgeist, the public has to get behind it. Beating the Soviets was worth it. People just signed up and said, it's worth it. I think that if we found out that a Mars base was crucial to protecting us from an asteroid storm or alien invasion, we would be there in 10 years.

So what I believe that NASA has to do is to embark on a massive campaign of disinformation and lies. [Loud laughter.] And I pledge to do whatever I can.

Discussion

JOHN GRUNSFELD: Welcome back to "Risk and Exploration." Our panelists are eager and ready to answer and discuss all of your questions and concerns. I imagine that the audience has plenty to offer, so I think what we will do is start. If you do ask a question or make a comment, please make sure and stand up, give your name and affiliation, and wait just a moment for them to cue up your microphone.

UNIDENTIFIED SPEAKER: I don't quite know how to phrase my question. It's been bothering me for the last day and a half. It was illustrated very nicely in the film clip we saw. There's an old set of characteristics of projects that has been going around for years, the last two of which involved praise and honor for the nonparticipants, and the last one is search for the guilty. The search for the guilty was illustrated there with the cross-examination of the astronaut for causing a failure. I guess I have been bothered for years by the fact that there always seems to be a need for institutional witch hunting. Somebody has to be guilty. This has got to be an inhibiting factor for managers and the people who have to make the tough decisions. I wonder how people feel about that. Is it really an inhibiting factor, or do you not think about it?

JOHN GRUNSFELD: That's a great question. I guess you missed the *Barcelona Times* in 1522 when Magellan didn't return, and they started the witch-hunt there. It's an interplay between Congress, the media, and the transparency that Graham discussed. We want to have a transparent space administration. That's part of our process.

JIM GARVIN: Sometimes the side effect of that mind-set is stimulated reexamination of programs. It has been said, and I've heard it said here, that we have programs in NASA. The one I speak for here, the Mars exploration program, continuously reinvents itself, precipitated by different types of catalysts. Sometimes they are the big setbacks. We did that after Mars Observer—the big witch-hunt of the early '90s. We did it after, as Steve said, the Climate Orbiter/Polar Lander issue, and we built a better program. So, that transparency and these effects you rarely see sometimes have positive consequences. How to live in the risk world without them in a highly visible public program is the debate we should all have. In the case of Mars, I can say that the level of incisiveness and the view that we took to do the rover, to do Mars Odyssey, which is still operating, and that we are applying to the Mars Reconnaissance Orbiter has been, perhaps, catalyzed by this mind-set.

GRAHAM YOST: In researching *From Earth to the Moon*, the feeling after the fire in Apollo I and the death of the crew was one of recommitment. The whole program just came together stronger and better than before out of that. I remember researching *Challenger*. There certainly was a witch-hunt, and Larry Malloy at Marshall became the fall guy. He once said that he understood that, as the middle manager in a corporation, his neck would be the one to go and his head would be the one to roll. I would say, and everyone here is in NASA after *Challenger* and you will see NASA after *Columbia* as well, that there is sort of a recommitment. I think, from the outside point of view, it's sort of a program that gets stronger.

STEVE SQUYRES: I think it serves no useful purpose nor is it in any way appropriate to have a search for the guilty parties. It just doesn't do anybody any good and should be avoided. At the same time, you cannot let your desire to avoid that scare you off from a ruthlessly self-critical evaluation of what went wrong. T. K. Mattingly last night said that every great success is preceded by failure. Certainly that was the case in the case of our mission, and I can tell you right now that the MER [Mars Exploration Rover] mission, as one simple example, would not have succeeded had it not been for the ruthlessly critical self-evaluation that NASA undertook of its Mars program, as Gene said, after the loss of MPL [Mars Polar Lander] and MCO [Mars Climate Orbiter]. I think the CAIB, the Columbia Accident Investigation Board, that Scott Hubbard was a part of, was a very necessary process. It was ruthlessly critical of the Agency in ways that were necessary and ways that will save lives in the future. We shouldn't have a witch-hunt, but you can't let it scare you off from doing a job you have to do when something goes wrong.

UNIDENTIFIED SPEAKER: I would like to add that I do picture myself sitting in Frank Borman's chair down here answering the questions. Now, you've said it was a good idea. You just gave up because you couldn't get the resources? Well, what kind of man are you, you know? It gives a person a little more courage to go tell the uncomfortable truth that you might have to tell sometimes.

UNIDENTIFIED SPEAKER: I think one of the questions that comes up as far as testing is the James Webb Space Telescope. The James Webb Space Telescope is a big telescope. That's why it has all that deployment, and it is still too small to do certain types of work that we know we would want to do at the end of the next decade: look for earth-type planets around nearby stars. I find that a compelling goal, and we need a bigger telescope. It will almost certainly be too big to test on planet Earth, and that gets back into the humans and robotic partnership. At what point are our goals important enough, our objectives well known, that the scale is such that we cannot test it on planet Earth? In the integrated test such that we might want to employ robots and/or humans as we do in the Hubble Space Telescope to check it out, how do you make that call? I know that this is something you've thought about, John.

UNIDENTIFIED SPEAKER: I would think that we have to do everything that we can with robots because they are probably quicker in most areas, and there will be some things they cannot do alone. When it comes to our dreams of big telescopes to find planets around other stars, I think we have to be really diligent in searching for ways to test them on the ground also. We just shouldn't give up on testing them on the ground, because I despair of convincing Congress that they have to fund us when we can't test it before we fly it. So, I think we have to be very, very imaginative about finding ways to test on the ground. We then still have to figure out that it's maybe not going to work.

There are some things we cannot adjust in the final stage and we cannot confirm on the ground. We have in mind flying constellations of telescopes that collect light from several different places and funnel it through a single combiner in the middle in an interferometer configuration, and with this method, you can build up the image sharpness that you would have from a telescope that is hundreds of meters, maybe kilometers or hundreds of kilometers across. You might want to do that to find out about those planets around other stars, but we just have no hope of testing that on the ground. Still, we must prove that it's going to work when you get there, so what are you going to use? Imagination? I couldn't tell you the answer today.

JIM GARVIN: We actually did experiments on the Moon with Apollo that you couldn't have tested on the Earth. The human beings, the crew set up some of the impressive arrays that we used to study the interior of the Moon and then experienced the collision of leftover space vehicles to generate a pulse; [that] was a novel, imaginative experiment that we did. I think there's an example of that. But there was something that we call Robotic Sample Return to bring back pieces of Mars to Earth. Some of us call it "Apollo without the astronauts" because of the complexity. The reason for that mission is because there are some things we think can only be done, at least until we reach projected technology state, with people in the loop. We either move a lot of people to Mars, and some of us would like to go, or we bring stuff back from Mars so the people here can work on it. Because of the testing limitations, you're there.

UNIDENTIFIED SPEAKER: One important point that I think is just good to get on tape is that there is no such thing as pure robotic exploration. The stuff comes back to the people who want to understand the science, so people are always involved. It's just a question of *where* are the people in proximity to the context. That's the evidence.

UNIDENTIFIED SPEAKER: Thank you, David Roberts. Problem for Jim Garvin and Steve Squyres and the panel. Besides Mars and Europa, what would be the likely planets or satellites for the next landers? Why? If possible, when?

UNIDENTIFIED SPEAKER: First, you left out the Moon. Going back to the Moon, it's a planet in its own right. While we visited tremendously with the humans, getting back there is a scientific and human operation proving ground. Mars is central to our vision, our implementation plan. So the Moon is a place. In fact, contrary to common belief, although not contrary to the science community, the Moon offers an interesting context for astrobiology. There's the question of early planetary crusts in which there may be aspects of our own history in the origins of life from which you gain context. Other than that, I would submit that it's a reconnaissance that will help answer that question. You named Europa. In January [2005], we will have the descent of the Huygens probe as part of the mission to Titan, unquestionably one of the most interesting objects in our solar system, certainly from the standpoint of planetary atmospheres and environments. Landing on that surface and sustaining landed experiments beyond the scope of Huygens is a wonderful step. We have a mission called Dawn that will visit two of the main belt asteroids, which are really planetary objects in their own right, Vesta and Ceres in this case, and landing on them, by virtue of what we find from the first nonlanded experiments that we'll be doing, I think is important. But I don't want to leave out Venus. Twelve impressive landings by our Soviet colleagues have left many questions that are so fundamental to understanding how big, rocky planets work, and their atmospheres, that we have no clue about. And, yet, that poses a risk challenge to sustain operations there robotically. I don't know whether the crews want to go yet, John, a little bit hard to get back in the gravity wall. There are a lot of places where landed experiments as a forerunner to sample returns and, ultimately, human landings, in my view, are important parts of our strategy.

STEVE SQUYRES: I think you answered it well. Actually, I'm glad you mentioned Venus because, you know, we have Venus, Earth, and Mars, and it brings us back to comments that Jean-Michel Cousteau and Sylvia Earle made this morning, which are, we like this planet, this is a nice planet we live on. And we also know we don't want to end up like Venus or Mars, and we don't understand any of them, including the Earth, all that well.

DAVID HALPERN: I'm happy somebody just mentioned Venus and Mars and the Earth. The point that was being made in the morning is not so much we don't want to end up like Venus with the hot house or the greenhouse gas—that's a separate issue. The simplest fact is that we know the topography—the ups and downs and the curves of Venus and Mars—to a much, much greater extent than we know the bottom of the ocean. That was one of the points that we were trying to make in the morning.

When it comes to exploration, if you want to explore, like Lewis and Clark, a continental area, what do you do? You first go there and you see what the height is and what the elevation is and where the streams are and where the gullies are. But we don't know that in the ocean. And the point that was being made—and then I think there was another comment made this afternoon—about 96 percent of the universe we don't know. Well, we don't even know 95 percent of the ocean. So, I just want to reiterate the point from this morning.

JOE FULLER: Joe Fuller, Futron Corporation. Right now, we're spending a lot of time, energy, resources, imaginations to reduce the risk associated with return to flight. And even though I'm sure everybody's going to do everything they can, there will still be risks. And if we think the unthinkable, what if there's another accident, you know, on the very next mission? It would have a devastating consequence. How do we get ahead of the curve in mitigating the risk of, you know, such a situation? I guess I'll refer to Graham Yost—he got very close to it in talking about, you know, dealing with the public.

GRAHAM YOST: Again, I think it comes back to the victim of success. The manned spaceflight in America has not been like test flight in America, where they had accidents all the time back in the '50s and people just kind of got used to that. God forbid, you know, that manned spaceflight had been like that. But it's hard to say what the public appetite is. Someone was asking me at the break about that, and I do maintain that the public is in many ways more concerned—I believe and I may be totally wrong in this—they're more concerned about the cost of things, because it's a pocketbook issue, than they are about the human risk. They're concerned about the human risk, but I think that they do feel that everyone is doing absolutely everything they can to make sure these people get back safely. I think that's just the tradition of the American spaceflight. And, so, I don't know what would happen to the Shuttle program. And, God forbid, you know, it took a long time, relatively, between *Challenger* and *Columbia*. That was a lot of flights, and it doesn't excuse it or make it okay, it's such an incredibly complicated machine—you all know. And the public doesn't know. But I think that the public accepts just the basic notion that it's risky.

JOHN GRUNSFELD: I don't know how Shannon feels, but I know for myself I'm more amazed each time we launch a Shuttle when you think of the tens of thousands of pieces, you know, that are checked out in the few seconds just prior to launch that all have to pass those checks before we actually leave. And the tens of thousands of people that all have to do their job just right before that Shuttle will leave the ground. It's always amazing to me that we do leave. I'm always a little surprised, when I'm in the vehicle, that we actually leave. I sort of prepare myself for that, you know, late countdown shutdown for some parameter out of limits, which many of us have experienced.

At the same time, Shannon, how do you compartmentalize? I know I do that. When I'm in the Shuttle thinking about the mission, I put the risk part of it, the scary part, in a little compartment, and it never really occurs to me when we're sitting out on the launch pad on four-and-a-half million pounds of explosives.

SHANNON LUCID: Well, you've made the decision that's what you're going to do and you've worked with the people that are doing everything. You've worked with the flight control team, you've worked with the Cape people, you know. And you know that they're doing the very best that they can. You know that they're only human; you know that mistakes can be made. And you've made your decision and that's what you're going to do. You don't sit there and analyze it and say, "What if?" at that time.

JOHN GRUNSFELD: That's right. In specific, we talk about the team aspects. And, as risk takers, regardless of whether it's earth, sea, or the stars, when you get into the vehicle or into the environment, ultimately, you have to trust those people who are making the decisions, and management, that they've done everything that they can. And often we talk about "as low as reasonably achievable" as a method of risk mitigation. For return to flight after *Columbia*, we have a very extensive guide, and the Space Operations Directorate has a very detailed return-to-flight plan. So, I think we're doing everything we can.

DAVID LONGNECKER: Hi, David Longnecker from the University of Pennsylvania. And my question is probably addressed to you, John. And that is, following up on what we just heard about mitigating risk, one risk we haven't really talked about so far—at least [to] any significant degree here—is the risk associated with a very large organization with multiple components, each doing their job to an optimal level, but, yet, creating a series of stovepipes that are not linked together across the organization. I'd be curious to know what NASA's doing to deal with that sort of linkage of risk across a huge organization.

JOHN GRUNSFELD: David, that's a great question. Thank you for asking that, and I think you saw on August 1st that NASA engaged in a rather large-scale transformation of its organization. The study of organizational risk in high-risk endeavors is a very mature study, but not very well understood. And you talked about stovepipes. The function of the transformation was to get NASA aligned behind a central goal. And you've heard that a lot. You have to have a clear goal. Everybody has to understand that goal, and everybody has to work together for that goal. In the Columbia Accident Investigation Board report, they talked about integration functions. And, so, one of the things that we've done is to strengthen, through consolidation and through this transformation, our ability to integrate. We have a Science Missions Directorate that now contains all of our science and has close ties with exploration systems, space operations, aeronautics; and we have an associate deputy administrator for integration who is the corporate conscience. And we develop our processes and policies to make sure that we have close integration between all these endeavors, whether it's the expendable launch vehicles in the science arena or human spaceflight and the exploration development. And part of that—and, I think, a key part of that—is so that we can incorporate lessons learned across the Agency into programs where those lessons may apply.

So, you know, that's not a full answer to your question, because it's only been a couple of months, but we're on the road towards trying to get that kind of integration and breaking down the stovepipes.

MITCH BARNEY: Mitch Barney, Goddard Space Flight Center. Ever since I was a college student, I've done my explorations in a bunny suit and clean-room booties.

I'm in the engineering side, developing new instrument-measurement techniques and technologies where failure is an option and the challenge is the risk—that's what brings you back day to day. Recently, the NASA environment for us has become a more competitive environment. We're competing and collaborating both internally and externally with private industry and with academia. And I wonder what the panel's response [is] to a question about the impact of a competitive environment on the risk that NASA's taking now. Dr. Mather and Dr. Garvin, you both mentioned the competitive aspects. So I wonder what you thought about competition and what it does—what's the impact on risk.

JOHN GRUNSFELD: I'd like to start this, if I could, with Steve Squyres. I came to NASA as a principal investigator in science programs and considered that process to be like swimming with sharks. Yeah, Steve?

STEVE SQUYRES: It is. I think that the competitive process that we go through—if exercised appropriately, if the selection process is done in an appropriate fashion—is one of the best risk mitigators that we have. My team, in various, different permutations and combinations, wrote three unsuccessful proposals—each of them, at the time, the best we could do, each of them with serious flaws—to agencies. We sent in the proposals. It was highly competitive. We lost. It was painful. We went back and we sharpened our pencils and we did a better job. And each proposal got better and better and, finally, on the fourth try, we managed to convince the Agency that something like MER was a good thing to do. I think that competitive process and the intensity of it—the pain and humiliation of losing a competition like that—drove us, and it drove us very hard to get better; and not just to write better proposals, but build safer, better, more-likely-to-succeed hardware.

And so, it's very important to have that competition. And I think the more broadly the net is cast—opening up the competition to industry, to universities large and small, across the Agency—to try to level the playing field so that everybody's competing on roughly equivalent terms, is a very, very valuable thing. And I dislike the competitive process intensely, but it's part of what has led to success of many programs doing that.

UNIDENTIFIED SPEAKER: I think the key for us at NASA is to make sure we provide you the tools and the ability to be able to compete head-to-head.

UNIDENTIFIED SPEAKER: If I may as well. I've seen the good and the bad of the competitive process for, in particular, science-driven experiments. And we haven't always gotten it right. I mean, Steve tells a good story, but we just spent the last couple of years going through that competitive process for 25 wonderful contests for missions to go to Mars, robotic missions, at this stage. And I think we've actually achieved a risk-based lesson learned from honing of that process. And I first saw that process as a loser, often, in the '90s, proposing instruments, but I later saw it from the standpoint of implementation, and I saw elegant things we want to do in space—both at Earth and on planets—get through the process

203

with great imagination and excitement, and then fail to sustain the cost envelope, the research envelope.

And I saw the community get smarter, that is, the integrated aerospace, university, NASA center, you know, community. And in this last Mars scout go-round—my one knothole in this—we had dozens of brilliant missions to Mars. And in the end, the final four—I'm not a basketball player—emerged after withering reviews by hundreds of individuals in which we spent more time worrying about being attentive to understanding risk. In fact, the most withering review is the risk of implementation review we do. And I think, you know, to some extent, it's the setbacks that have honed that. So that I'm much more confident from all these analyses that we can do these things. Now, the question is, I think, how do we maintain continuous improvement of that process when we reach a certain level of performance—success from MER, success from COBE?

JOHN GRUNSFELD: I'm thinking that there are a lot of rules of the game that govern how it all plays out and that at NASA Headquarters, when we set up the rules of engagement, we basically determine the outcome—in a way that we may not anticipate—of how organizations grow or die. And the ability of organizations to grow or die as a response to the competitive process is part of capitalism, and it's part of the sort of basic religion of America, practically. But it does have some unintended consequences. Creative destruction is sort of [a] motto that people carry, and a lot of us may lose. So, well, that's just part of the deal. And I think that we need those competitive forces and, certainly, I've had a lot of losing proposals as well, so they know that and they deserve to lose, but maybe next time.

STEVE SQUYRES: There's one other aspect that we've mostly been doing with the science side, because we're a science-heavy panel, but Jim brought up the metrics by which you decide who wins the proposal. In the science case, we want projects that are viable, scientifically top-notch, and so on. One of the duties of the commander on a Space Shuttle flight—and I presume it was true of a Saturn V flight as well—is to remind the new fliers of the group that they're launching on the lowest bidder's successful project. So sometimes those measures of effectiveness may be at odds with low risk.

JOHN GRUNSFELD: It must be getting late. Well, Jim, let me pick on you for a minute and ask you something. You've flown a Shuttle experiment, and I've thought about this. My Ph.D. thesis was on a Space Shuttle. It was on the Space Shuttle *Challenger*, the flight before the fateful one, and I didn't think about the risk element to the crew at all. When you flew your laser experiment, did you think at all about the risk the crew was taking to get science for you and your colleagues?

JIM GARVIN: Well, in fact, John, we did, and for two reasons: One was the risk that we were afraid we were imposing on them. We had 45 millijoules, the number in laser metrics, a non-eye-safe, infrared laser transmitter. The light could have blinded the crew looking out the back window. And so we developed flight rules and procedures with the crew so that they would be sleeping, often, when we operated. For two reasons: We didn't want them moving around, because we were trying to measure little things on there, and, second, we didn't want that risk. But I remember vividly the launch of STS-72, which was in January of '96, and the biggest snowstorm in a decade in Washington[, DC].

And I remember thinking that the crew were launching, you know, not really thinking about the weather or getting to the payload operation control center at Goddard where they were running the experiment, which was causing me great stress, but I thought, they were riding these seven million pounds of thrust to carry our team's hundred kilos of stuff to try out an idea. And I thought how lucky we were. Because we had all the infrastructure that got them there into orbit successfully—in this case, it was a recovery mission for the Japanese—but also, to let us have this window on the world with this flight. We went with checked and set parameters, so I thought, if our straw is the one that breaks the back for the crew and, also for the mission and the Shuttle, that would be, you know, a tremendous setback.

We were scheduled to have an experiment like this on the first Shuttle launch for science out of Vandenberg. So, we were to go into polar orbit with the Shuttle to do experiments looking over the polar ice, being a big thing we wanted to measure. But, at any rate, it took us nine years to get back to our experiment on *Endeavour* when we flew. I just think those are the challenges of human spaceflight.

UNIDENTIFIED SPEAKER: I know, from the risk-taker's side, that the decision to go is very easy when it's making great science or great exploration. That makes it a much easier discussion to think about the risks versus the rewards. There are folks I've flown with out here in the audience and we've done tremendous science.

JOHN GRUNSFELD: Any more? Going once, twice. Very good. Well, I want to thank all of the speakers. Let's give them all a great big hand. I also want to take the time, once again, to thank the Naval Postgraduate School. It's been a tremendous venue for us and a great environment to have these discussions today for "Sea and Stars." Also for Ames, which has helped facilitate this. Don't forget, tonight is dinner and a movie. We'll be watching the *Endurance*. And for those out in the listening world, thanks for watching NASA TV.

205

why we explore

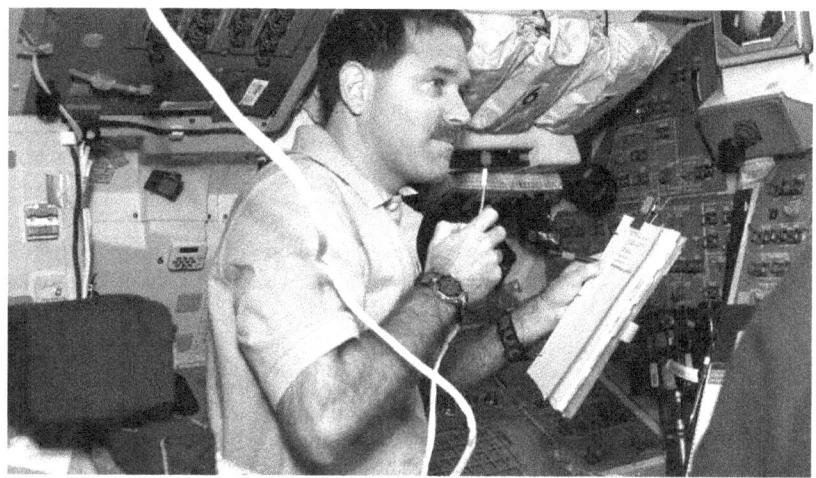

Why We Explore

"Welcome to Risk and Exploration —Earth, Sea, and the Stars." Today's session is entitled "Why We Explore," but I'm hoping that, mostly, we can make it a dialogue, up close and personal. I'm John Grunsfeld. I'm the NASA chief scientist and an astronaut.

I think we have started getting into the discussions on risk and exploration, into some of the thorny questions about how do we make decisions. How do we use our judgment? How do we, as institutional managers of a public institution, make decisions on behalf of the American people, and with oversight of the Congress, that can stand the test of time, without being so risk averse that we don't do anything interesting?

There's a couple of things I'd like to show this morning that are personal, that are professional as chief scientist, and then, representing the Agency, and then, looking forward.

I think we'd be remiss in all of this discussion if we avoided the topic of why we're not sending a Space Shuttle back to the Hubble to service it. So I'll address that in a second.

One of our favorite cartoons shows a Conestoga wagon heading across the Great Plains. And the title reads "Alarmed by the many dangers, the pioneers abandoned westward

John Grunsfeld
NASA Chief Scientist and Astronaut

Astronaut, astrophysicist, and mountaineer, John M. Grunsfeld served as NASA's chief scientist from 2003–2004. Grunsfeld is a veteran of four Space Shuttle flights. In 1999 and 2002 he took part in a total of five successful spacewalks to upgrade the Hubble Space Telescope.

A native of Chicago, Grunsfeld received a bachelor's degree in physics from the Massachusetts Institute of Technology in 1980. He earned a masters degree and a doctorate in physics from the University of Chicago in 1984 and 1988, respectively. Grunsfeld was selected as a NASA astronaut in 1992. His first flight assignment came in 1995 on board the Space Shuttle *Endeavour* on STS-67. In 1997, Grunsfeld served as flight engineer for the Space Shuttle *Atlantis* during STS-81 and a 10-day mission to Russia's *Mir* space station. He has logged over 45 days in space, including 37 hours and 32 minutes working outside the Space Shuttle.

exploration except for a series of unmanned prairie-probe vehicles." You know, I think many people have summed up succinctly why humans explore. Because we want to go. In the face of danger, but managed risk. I am absolutely positive that our outward expansion from the cradle that planet Earth is, will not be one of strictly unmanned probes, but we will be heading out across the prairies. Why [do] we explore? Right now, Spirit at Columbia Hills [Mars] is poised to look over those hills and see what's beyond.

I'm an explorer who is trained by a group called the National Outdoor Leadership School, and we're privileged to have John Gans, the director of the National Outdoor Leadership School, here. I went there in high school and it was to learn to be a better risk manager, a better leader in the outdoors and, hopefully, not to be reckless like most teenagers.

But my interest in exploration was largely driven through the pages of *National Geographic*, through the movies of Jacques Cousteau, while growing up on the south side of Chicago, that I was able to explore vicariously. But I wanted to go. I have a passion for exploration, and I have a weakness. When I see something like Columbia Hills, I have a need to look over that hill. And it's a real challenge.

It's a real challenge because you set limits for yourself. And as mountaineers, we set limits for ourselves. We have to summit by a certain time so we can make it back safely. And I'm constantly torn, wanting to go further, especially when I'm on professional travel and I take a day off to go hike. I say, "Well, I only have one day and I'll go this far." And I get that far and I look forward and I say, "Boy, I've got to go a little further."

So, that's what we're doing with Spirit and Opportunity on Mars now. We have the opportunity to go further because the rovers are still running, they're still doing great. You know, we had a 90- or a 120-day mission and we're well beyond that now, and we have hope they can go much further.

This May and June, I had the opportunity to try and climb a little hill in Alaska called Denali—Mt. McKinley. It's 6,157 meters, 20,320 feet tall. This is a serious expedition. It's not quite the kind of thing that Ed Viesturs does, but it's, I think, comparable in many ways.

It's at 63 degrees north latitude. That makes it perhaps the coldest mountain on planet Earth. You start out already basically in the Arctic. Its conditions on the summit are comparable to Everest in winter. The Alaska Range is a large landmass that extends up out of plains, basically, a few hundred feet in altitude. It sees the full brunt of arctic weather. And, so, it seemed like an appropriate challenge.

Now, in order to do this as an astronaut—and I see Colonel Cabana in the audience—this was my third try. The first time I tried, as an astronaut, I felt compelled to write a mission statement and a risk-mitigation statement that I submitted to my boss, Colonel Cabana, then chief of the astronaut office, so that I could get permission to go, so to speak. Even though it was personal leave.

That's the way I view risk management on this climb: you have my crew notebook with checklists. And I think I'm the only mountaineer I know who goes up with checklists and says, "Okay." And part of that was, I recognized that at high altitude I will be hypoxic, I will make mistakes. And this was one of my mechanisms to prevent myself from making mistakes. I still made lots of mistakes. I think back now and I think, "It's in the checklist; how could I have missed it?"

But it's one mechanism of risk mitigation that we use very often in the space business, because the line between life and death is so fine. We heard in this talk something that I think is very characteristic: the farther you go from base camp, if the smiles get bigger, you have the right team. And we lived, basically, on a glacier for 23 days. All of our water came from melting snow.

Just a great experience. No cell phones, no beepers, no Blackberrys, really just existing in a very primal way, but with the aid of high technology, and that's something I think is part of the real spirit of exploration, that trying to go to the next hill. And I got up to the top and I looked back and I waved at Dave Schuman, who is another NASA Headquarters employee. I said, "Dave," and I had to yell.

I HAVE A PASSION FOR EXPLORATION, AND I HAVE A WEAKNESS. WHEN I SEE SOMETHING LIKE COLUMBIA HILLS, I HAVE A NEED TO LOOK OVER THAT HILL. AND IT'S A REAL CHALLENGE.

I said, "Dave, I have bad news." And he was thinking "Oh, no." We thought we were close to the summit—we'd been on the summit ridge for about two hours climbing up from something called the "Football Field." And it's tedious. What Ed Viesturs said is right. You take a step, you breathe a bunch of times, and you take a step. And every time you stop to breathe, you look forward to see how much longer it is. And, very often, you don't see the top, you know, [you have to] climb another ridge. And I said, "I have bad news. There's no place else to go but down."

I was actually worried about sort of an anticlimactic feeling. This was my third try, and I just couldn't believe I was actually standing on the highest point in North America—just an unbelievable feeling. I was half laughing— my climbing buddies say hysterically—and half crying. I just couldn't believe it. So we had three NASA employees on the summit of North America on June 7th of this year, 91 years after the first ascent.

A lot of people have climbed to the top of this mountain—about 12,000. About one out of a hundred perish in this. My risk management plan was to go through a book called *Accidents in North American Mountaineering*. It's published every year. Just the fact that a book like this is published means that mountaineers are very sensitive to this issue of risk and that we try and learn from others' mistakes.

I went through basically every mountaineering accident on Mt. McKinley and in the Alaska Range from 1969 to about 1992 and came up with common causes, behaviors that led to those accidents, and then asked myself, how can I avoid those behaviors? And, so, that was also in my little notebook. And I'd review that every night and then review it with the team. "Okay, we're not going to do this. We're always going to stay roped, no matter where we are. We're always going to carry an ice axe."

We pretty much beat it away so that if you do the statistics, it became more like 1 in 10,000. And one of the things that people think about mountaineering is that it's high-risk behavior. In fact, a mountaineer who climbs recreationally, as I do, is about three times more likely to die from heart disease in the United States than from a mountaineering accident. But, of course, when there is a mountaineering accident, and a rescue, that short-term drama that we discussed here is what plays big, and not the many, many safe expeditions.

The other thing that we will talk about in the discussion is the Hubble decision. On January 16th [2004], Sean O'Keefe, the Administrator of NASA, Ed Weiler, and I went out to the Goddard Space Flight Center to announce to the Hubbard Space Telescope servicing team that Mr. O'Keefe had made the hard decision that we were not going to return to the Hubble Space Telescope for a fifth servicing mission with the Space Shuttle.

This hit me extremely hard. I am literally a "Hubble Hugger," as I think many of you know. I've had the privilege of visiting the Hubble Space Telescope twice. I'm a professional astronomer. I know Bob Parker is here somewhere, he's another astronomer astronaut, and I'm sure he can appreciate how tough this was. But Mr. O'Keefe looked at all the elements post-*Columbia*, and, in fact, our last mission was on *Columbia* up to the Hubble in March of 2002. And he looked at the recommendations of the Columbia Accident Investigation Board.

Hubble has a clock, an internal clock. And that clock is driven by gyroscopes and batteries. And sometime in the next two to three years, the gyroscopes that are on Hubble will wear out, and Hubble won't be able to do science anymore. Not too much longer after that, the batteries will run out of juice, their ability to charge and recharge, and at that point, the telescope will go cold and won't be able to be recovered.

So we have to get to Hubble before the batteries die. And if you look at the recommendations of the Columbia Accident Investigation Board, and if you say we're going to satisfy every single recommendation before we go to flight and you say that we're not going to succumb to schedule pressure again, then when you look at the risk-to-benefit of using a Space Shuttle, you put yourself in a real box.

One of the boxes goes like the following: Imagine that we press forward with a Hubble servicing mission with the Shuttle. We have the crew trained, we have the big team trained, and we're on the pad. You know, maybe even we have liquid oxygen boiling off and the hissing and the moaning. And, in the launch count, we find out that something's not working right. A computer is down, a multiplexer isn't working, some communication link on the ground isn't working. Our flight rules would say, "Don't launch."

But whoever is in the hot seat that day will feel enormous schedule pressure to launch that mission anyway, because Hubble won't wait. We're all success-oriented, that's what we drive to. And Mr. O'Keefe didn't want to put any manager in that position.

Worse, when we go to the Hubble orbit, we launch due east. And, so, the only self-rescue capability we have—and that's another very important element in mountaineering or any outdoor adventure or going down in caves and, certainly, in the Antarctic—is limited to what you really have on the Shuttle. And, so, early in his analysis, he said if we're going to go to Hubble, we want to have a second Space Shuttle available on the pad so that you could launch within less than 30 days, which is probably the maximum you could keep a Shuttle crew going in orbit, in case of a *Columbia*-like accident. Well, imagine the enormous pressure if you had to execute that—of the second Shuttle to go rescue the crew.

Would we do it? Of course we would. If we put ourselves in that position, we would do everything we could to mount that rescue mission. And the same thing if the weather's not good, if something's wrong with that second Shuttle. And, I think, about half the time, there's some issue that delays us. We're getting better and better. I know two of my four missions have been delayed by a number of months. Many other missions have been delayed even as close to a few seconds prior to launch, when an engine will shut down for good reason, and we then recycle to two or three months later.

That's not acceptable if we're doing a rescue mission, even if it is a best-effort rescue mission. So I think the managers would feel that extreme schedule pressure that would put another crew at risk. So Mr. O'Keefe just felt that, as the top banana risk manager for the Agency, he didn't want to put us in that position.

That's a tough call. We all love Hubble. Hubble does extremely important science and is, perhaps, the most important scientific instrument ever created by humans. So this hit many of us hard, and it's that emotional side that makes risk-analysis and decision-making so hard. Someone said that the decisions we don't have control over are the ones that we worry the most about, and the ones we do have control over, we worry the least about. Well, this is one that I know Mr. O'Keefe has worried the most about.

And it really is compliance with all of the recommendations of the Columbia Accident Investigation Board and where we have raised the bar to make sure that we fly safely with the Shuttle—as safely as we can.

What we also heard in this conference [is] that the only limit is our imagination. Absolutely true. So the question you have to ask—and I did ask the Administrator—"Okay, if we can't go back with the Shuttle to service Hubble, how can we service Hubble?" I didn't quite put it that way, but I came back and said, "If we can service Hubble without the Shuttle, can we go forward with that?"

And I explained to him that it might be possible to use a robot to service Hubble. Now, keep in mind that I'm proposing something that puts me, as a spacewalking astronaut, out of business. But that's exactly what we want to do. We want to take routine operations of servicing—things that we can do with robots, things that we pioneered using humans that now robotics can do—and replace humans in hazardous situations.

EVA, Extravehicular Activity, is a very hazardous activity. We've been very fortunate in our spacewalks and there have been some close calls. That being said, servicing Hubble robotically will be a true, high-performance challenge. So, it's not clear that we can do it yet. But Mr. O'Keefe said we can go investigate that. This was an idea that came out of the extremely talented team at the Goddard Space Flight Center led by Frank Cepollina, one of our top inventors and out-of-the-box thinkers, a true explorer, one of the people responsible for the first servicing mission.

Remember, Hubble was a "space turkey," a "dog in space," "space junk." All of those things that we heard after it was launched, just because the mirror was ground to the wrong shape. Now, it was actually the best mirror ever created, but it was the wrong shape. Well, we went up and put contact lenses on it, corrective optics, on the first servicing mission. And for three years, people [had] said, "You can't do it. People can't service it. It won't work. It will be too hard. You'll end up destroying the telescope." But we did it.

It was that same team that came forward and said, "We think we might be able to send robots to the rescue." Well, that alone wasn't quite enough to put us collectively over the edge to suggest that we actually should proceed with the robotic mission, until we started listing the key technologies that we would have to prove to be able to service Hubble. And those technologies were: autonomous robotic rendezvous with a spacecraft, proximity operations close to the telescope, reaching out and grabbing the telescope, effectively a docking; doing an assembly, putting a new spacecraft underneath the Hubble—robotic assembly, and then, having dexterous robotics, agile robotics that can feel, to be able to service the telescope the way humans do, new tool development.

We looked at that list and I said, "Boy, that list looks exactly like our top list of things we need to learn how to do to explore—to go to the Moon, Mars, and beyond." And so the idea came up of using Hubble to be a catalyst for exploration.

Because, after all, what is Hubble? Hubble is out exploring the universe. It's our eyes for exploring the universe. It already is doing our exploration mission.

And what a great part of the Hubble story that it can continue to do science and be that spark that allows us to go further on.

Hubble is very hard to work on. This is going to be pushing what we've done in space, you know, maybe one or two generations. But people perform at their highest when we give them high-performance challenges.

If you ask people the easy things, they'll do it, but when you ask people who are passionate to do something hard, they'll do it well, and they'll pull out all the stops. And we've heard that in all of our panelists and in contributions from the audience.

Once we installed the Advanced Camera for Surveys in the telescope, there are a lot of very delicate operations. One of the good things about a robot is once its done the operation once, it can repeat it over and over again. And we have a greater knowledge of the metrology—of all of the measurements of Hubble—than any spacecraft ever, due to four servicing missions and all of the metrology we did on the ground—all of the measurements we made. So that we could build instruments on the ground while Hubble's in space and know that they'll fit. So this is the best setup we're probably going to have to try some of these hard things.

The proposed robotic servicing mission will launch on an expendable vehicle. We've got to get it up there sometime around late 2007 or 2008, so those of you who've heard about the MER, you know, 34 months wasn't long enough; well, we have about the same amount of time. It'll have two parts. It'll have a part that's going to stay with the telescope and one that will leave. Once it's on orbit, a robotic arm, much like the Space Shuttle, will be deployed. And there'll be people involved—this is not a push a button and it goes. Folks on the ground will be monitoring this and, maybe, controlling it.

We're going to grab Hubble in exactly the same way we did with the Shuttle. We'll have the same end effector—a very similar arm, a similar approach—and we'll use the same spots on the Hubble that we grab with the Shuttle. So, we're still in known territory, we're just using a robot. The robot will then put itself on the bottom. And that's exactly what we do with the Shuttle—we grab Hubble and we put the Shuttle underneath. And we latch with these exact same latches. Well, now the robot's doing it.

Once we let go, now we have to get the arms and the hands. To do that, we're going to use this special-purpose dexterous manipulator. It's already flight-ready. It's a Canadian arm called "Dextre"—that's the call sign. And it was built to service electronics on the International Space Station.

Well, we're going to steal that first and use it on Hubble. It'll deploy some cables to hook up the new spacecraft into the Hubble. This is something we feel comfortable doing, except for the part with the connectors. The connectors are always tricky—they're tricky for people with hands, especially when you're wearing these big space gloves. In two weeks, we built prototype tools that were able to take these connectors on the ground-based version on and off. And so we're reasonably confident we'll be able to do this.

215

But of course, we have to expect the unexpected, and Hubble always provides unexpected surprises. So we're going to involve all of the spacewalking astronauts and folks who have controlled robots on Hubble to go through all of these and think what could go wrong, and make sure we have the robot designed to do that.

Next, we're going to take out the wide-field camera. There's just two bolts and a ground strap and it pulls out externally. And we're going to stow the old one and put a new one in. The nice thing is, once we've taken it out, the robot now knows every motion it takes to put the new one back in. And we'll have extra cameras. One of the things about this robot is it can actually feel force. We'll have monitoring so that if it hits something, we see the force on that particular joint rising, and we can back off a little bit and change the attitude.

We also have to hook the new spacecraft into the brains of Hubble. That'll be another connector. It will be on the computer that Mike Foale put in. There's a connector on the top, fortunately, not on the side, that's just a shorting plug. So, take it off and put a cable on and then close the door on the cable.

Now we get into the really hard stuff, which is to take the corrective optics out. They are not needed anymore. All new instruments have the corrective optics built in. These are the sides of the refrigerator, so this is doing the job of Jim Newman, who had that advance camera and put in the Cosmic Origins Spectrograph. Once that's done, the servicing part goes away. Hubble, hopefully, will get between three and eight years of extended life, and then, at the end of life, we have to safely deorbit Hubble. Again, it's a safety and a risk issue. There is about a 1:250 chance that some large part of Hubble will survive to the ground in a populated area, and that risk is just too high.

So that little package on the bottom that has the new batteries in it also has deorbit engines. So Hubble will deploy all of its booms, start charging up the new batteries, do its science, and then sometime, perhaps as late as 2015, we'll feather the arrays, much like Mike Melvill feathered the wings on *SpaceShipOne*, and fire the deorbit module.

I'm hoping to be on a cruise ship somewhere in the Pacific to watch Hubble fly over and reenter. I think we'll all have to have a big party and really celebrate an incredible voyage. It will be this voyage that will have helped stimulate and advance us, probably by five or six years, in the exploration effort.

So that's the plan for the robotics. We actually have some contracts in place now. I think it was last Friday we announced that Lockheed Martin had won the contract to build part of the spacecraft. It is going to be assembled at the Goddard Space Flight Center as an in-house project. People say three years is impossible. But the really good news is that we have a tremendous amount of hardware already built, because we were on the road to a Servicing Mission 4.

Hubble has produced great images. How many people have seen the Hubble Deep Field? Or the Ultra Deep Field? An amazing picture. A thousand galaxies. It took 11 days staring into a blank part of space. If you hold a soda straw up to the sky at night and look through it, that's about the area of the picture. If you look at what's in the background in this relatively short exposure with this new camera, you see a lot of things. If you add up all of the spots, each of which is another galaxy, there are six thousand galaxies. Remember, that's just a little soda straw with a relatively short exposure. Each one of those galaxies has 100 to 200 billion stars. You heard me say yesterday, 10 years ago we didn't know about any other planets outside of our own solar system. Now we know, in just the nearby stars, of over 125. There are about twice that many that are being investigated to be confirmed. We now think planetary systems are common. So you can add it up just in this picture alone: 6 thousand times 200 billion times a couple of planets per star. There are a lot of planets out there. It's pretty mind-boggling.

That's where we're going on Hubble. Where are we going next with the Space Shuttle? Well, we're going back to the Space Station. In the President's vision for space exploration, our first task is to return the Shuttle safely to flight. The team is working through that. They are working through it with a passion as well. It is also hard.

I HOPE TODAY IS THE START OF A DIALOGUE THAT YOU ALL HAVE

WITH US AND THAT WE HAVE WITH YOU. . . .

We are finding a lot of challenges, not the least of which is we don't really have a Shuttle we can launch to test the new foam changes. So, we have our best and brightest engineers working on it. We have Admiral Cantrell helping us with the safety issues. We have really pulled out all the stops.

The crew that is going to go is led by Eileen Collins and piloted by Jim Kelly, both very experienced. I flew with Wendy Lawrence on my first mission. She's an incredibly hard worker, intense and talented, from the United States Navy. Charlie Camarda, Andy Thomas, Steve Robinson, Soichi Noguchi from JAXA, the Japanese Space Agency. A really exciting crew. We are going to dock with the Space Station. We are going to evaluate our techniques for inspecting the orbiter and for repairing the orbiter with some EVA flight tests.

The crew patch [referencing presentation slide] has the crew names around the outside and on the orbit. It has this swoosh and the STS-107 outline from their crew patch in recognition of that crew.

Before you leave today, we have STS-114 pins for every one of you. I want you to wear those as a reminder of this conference and the work we have ahead of us. There are going to be hundreds of risk decisions, reward considerations, and judgments that we are going to make before we return to flight.

I hope today is the start of a dialogue that you all have with us and that we have with you, and that you will take with you to the groups with whom you work, whether it's within NASA or outside of NASA. To continue this

217

discussion, either as organizations or as smaller groups or as individuals, we can come back and keep this an ongoing dialogue. I think this continuing dialogue is very important if we are going to minimize the risk and maximize the return in our endeavors. STS-114 is part of that dialogue, and you will all be part of it now—those of you within NASA, by definition, and those outside—because of your participation here. We thank you for that.

As [NASA] Chief Scientist, I get to spend a little time in the House Science Committee room in Congress, probably more than I'd like. There are some things written on the wall that I think are really fantastic, and every time I sit there, thinking, What am I gonna say? or What are they gonna ask me?, I look up on the wall and read, "For I dipped into the Future, far as human eye could see; saw the vision of [new] worlds, and all the wonder that would be." That's from Tennyson. Again, this is something that I think drives us all.

Discussion

JOHN GRUNSFELD: With that, I would like to really open the floor completely to any discussion we might have, to talk about some of the overarching themes that came out of this meeting, anything we haven't covered. I want this to be very personal, and if folks don't ask questions, and even if they do, I'm going to ask folks that haven't participated yet to volunteer. In fact, let me try and stimulate that a little bit.

Why don't I ask our moderators if I can put them on the spot: Dave Halpern, Chris McKay, and our dinner speaker, Mike Foale, put you on the spot. We'll just start talking a little bit.

I think it was clear to me that there was general agreement that the greatest risk is not to explore at all. I think that is something that we have to communicate to folks. That we should not get so risk-averse that we just don't go out and explore. I have covered this one, but also the greatest risk is the lack of imagination. There are a lot of "greatest risks." They are all up there as the pinnacle of greatest risks. I had written down in my notes: Always expect the unexpected. Just like being asked to come down and talk. The other one, which I really liked, was Miles O'Brien saying, "The public is not as wimpy as we think." I can see no greater example of the public flying a new space ship than *SpaceShipOne* this morning.

Last year in the United States, about 40,000 people died in car accidents, of which 22,000 were not wearing seatbelts. I call that stupid. Folks who do that aren't thinking about the risk and consequence. In exploration, it is harder than that. Mike Melvill, on his first flight, and, as far as I know, on this flight, had no pressure suit. When you get above 50,000 feet, the remaining atmospheric pressure is such that the partial pressure of carbon dioxide and water vapor in your lungs dominates, and oxygen cannot get into your blood. Water at body temperature at low pressures will start spontaneously boiling if you expose it to a vacuum.

Mike Melvill had no pressure suit. That's hanging it out. Why? Why wasn't he wearing a seatbelt and a pressure suit? Well, performance. They are at the edge of the performance of what we can do with a vehicle like that, and that's the decision they made to take that risk. It paid off.

They have a very, very good safety program. They look at all of the considerations, and they consider their pressure vessel to be one that can't fail, at least given the risks that they are willing to take at this stage. So they are true pioneers.

In aviation, so many pilots died early on. As a result of that, we developed lots of the safety mechanisms that test pilots like Mike Melvill use to fly in *SpaceShipOne*. This risk-judgment-benefit is really tough stuff, and I think it's great when it succeeds. We need to keep pushing.

Miles also said, "Exploration is driven by fear, greed and curiosity." I like to stay on the curiosity end of things. Focus on the target. Get the right target. Have an overarching shared goal. It is just wonderful now that NASA has that in the vision for space exploration. We need to keep focusing on that. The other thing that I thought was interesting is that 96 percent of the energy content of the universe is in stuff that we have absolutely no clue what it is. I think that's great.

What I didn't know is 96 percent of the undersea environment has not been explored. Folks say the easy targets have been explored, and it gets much harder from here. Largely, that's true. Technology is no substitute for experience and leadership.

I am going to put Mike Foale on the spot right away, because I've talked a little bit about NOLS [National Outdoor Leadership School], and technology is no substitute for experience and leadership. Mike, in the Astronaut Corps now we are doing some things to try and enhance our expedition leadership. Maybe you can talk about just a couple of those things, like the NOLS, like the NEEMO [NASA Extreme Environment Mission Operations], and let folks know what we're doing.

MICHAEL FOALE: Well, John, I covered a bit of that in my dinner talk. To carry out any technologically advanced exploration mission which involves complicated techniques and new equipment, you have to be trained. You have to know how to use your equipment. If you take really fancy gear with you and you don't know how to use it and you waste time on it in a blizzard or on top of a mountain, you may actually end up risking more by messing with it than leaving it behind. When we train, we are basically putting aside that risk. We are mitigating the risk of carrying out the mission when we actually, finally, get to space. With astronaut training, just as the training we heard about going under the sea or into the deep caves, certainly you train for mountain climbing, we try and train so that we will perform better once we are there.

However, the training is not necessarily without risk. The training is not necessarily in the simulator, where I think you're pretty safe, unless a brown recluse bites you, which happened with Joe Engle once. Some of our training involves going outdoors. Some of it involves flying in aircraft that could have a malfunction, in particular, as we look toward exploration beyond Earth orbit, and in particular, when we can't turn around. Think of the Apollo 13 example, when your problem occurs on the way out to your destination, or where you are forced to go a long way out before you get to come back, then you are in a situation of survival or making the best of all the materials and resources at hand.

We have started to develop training that will take astronauts in small groups, crew size roughly. Six is what we've been choosing. We are then working with the National Outdoor Leadership School for one, the Canadian government for another, and the NEEMO group in Key Largo, Florida, to explore and develop leadership, followership, and self-management skills in our crews. That is one task.

Second, we are starting to get a feel for what we need to do when we are isolated and when we are dependent only on each other and the things we have at hand. That training we are still in the process of discussing and inventing.

I showed you three cases. One is the National Outdoor Leadership School, where we go out either into the Canyon lands or into a mountain environment. Another one has been exercises. We have taken part actually starting off in Cold Lake, Canada, where basically we are given a scenario more of "expeditioning" from one point to another. It is kind of a fancy series of walking through the snow, managing yourselves, looking after your team, and managing the equipment that you have with you. It is led by instructors who already have a plan on what the exercise is going to be, and it is covered in terms of risk, because we also have the full resources of the Canadian Armed Forces to get us out of there if we actually got seriously hurt.

Another analogue that we have been exploring and using is the Aquarius Laboratory in Key Largo, Florida. There we've put together crews that are three astronauts with three nonastronauts. We have actually included our Mission Control flight leads, who would normally be in the Control Center controlling a mission, to take part in those dives and those missions. And there we have actually solved another issue which is the classic problem of "What on Earth is ground control thinking!" when you get these strange instructions. Especially if you actually have people who have a stake in your activities. Scientists who are not in the team but [are] back at home in safe conditions will be asking things of you that might be rather difficult or seem rather strange or irrelevant at that point in your difficulties. And we're trying to bring together the Mission Control teams with the astronaut teams that would be deployed so that they would see each other's problems. And we've done that two or three times now.

Looking to the future, as we plan for moving to the Moon first and then Mars, we need to develop further the idea of being able to maintain our equipment, look after those resources that we have, even if they break and are a long, long time away from any kind of refurbishment back on Earth. To that extent, we want to actually follow-up on some of the Apollo lessons that we saw to teach geology at first. And these were in the deserts, I think always in the United States. Because of the Martian meteorite interest now, it's very exciting. We have thought about attempting expeditions, taking part with other scientific expeditions where there is a scientific goal that we, the astronaut office, do not have a stake in but they have a stake in our performance. So, that would be realistic and an analogue to a real mission on Mars or the Moon, where we have to carry out some of the grunt work—the deployed work—that would be required.

We are also looking at going to hot deserts, where meteorites also occur. This would combine the expedition training with acquiring field geology skills that would also be required of a small group going to the Moon or Mars. I think that summarizes it.

I should just add, once we get to these deployed situations, there are risks. And I myself was a little perplexed a number of years ago, when I was asked to come up with the whole safety/risk mitigation plan. And I had to have a safety review of what we were talking about and planning. At that time, we were sending roughly six astronauts at a time for survival training in Russia. And at that time, I talked about needing to have some insight into the other partner's processes. The same would go for the Canadians. The same would go if we hire another group, such as NOLS, for example. NASA has to know what you do when we use your services.

With Russia, they had had a number of helicopter accidents, and the training we were proposing was to do some of those search and rescue exercises with the Russian helicopters. And in the end, people were so just alarmed by the stories coming out of Russia we had to turn that off. I didn't know how to fill out a kind of safety/risk plan or matrix on that. Sometimes, you just have to use good judgment. You use your intuition. You ask all the questions that you can. We heard how James Cameron manages his film set. That was interesting to me, because he doesn't have the formal process that NASA has in its bureaucracy, forced on as a result of many, many mishaps over many, many years of experience. Sometimes, in smaller groups, you have to use judgment. And I think we are going to be in a position of having to use our judgment as we assess some of these new activities, not only just processes and safety reviews.

JOHN GRUNSFELD: One of the things that you brought up was working with our international partners. That, I'm sure, is going to be a major issue with pushing out beyond lower Earth orbit to the Moon, Mars, and beyond. And it happens on Earth frequently, the issue of different international cultures. And we heard a little bit of that. I think that's one we're really going to have to grapple with. Not just the day-to-day living types of things, but how different cultures deal with the safety issue. Whether it's documented or whether it's trusting people's good judgment.

MIKE FOALE: I know Chris, and certainly others from Ames who have been doing research, they've had to work, in particular, with Russians. And I know James Cameron did. I'd love to know what their opinions are and how they manage insight into systems that they don't know all about.

CHRIS MCKAY: It's difficult, and in Russia, in particular helicopters flights—flying with the U.S. Navy in Antarctica is very different than flying with Aeroflot in Siberia. And we'd like to take you on both trips, Mike. Actually, we would like to get you into the ice-covered lakes. I think you and Dale and I ought to talk after. But, actually, what I would like to do is come back to this conference and just think about it a little bit. It's been an incredibly fun and interesting conference. I can't remember when I've enjoyed one as much. My question to the audience is, how do we make it a useful conference? How do we take what we stated here and make it something that could be useful—and I would say, useful to whom? Useful to three parts of NASA. The robotics program, which, after we heard Steve Squyres talk, I think is clearly in need of a better risk assessment strategy. The near-term human program, which we're hearing about. I think John, Mike, and others are well involved in that, and I think that's been in the impetus for this conference. And I applaud them for their efforts. And I think that's the light that makes a way of doing something useful possible, [your] attention to this. And, then, third is, [as] Mike was just saying, long-term human exploration of distant planets—which is going to be a completely different category of risk and danger. So, how do we make this conference useful to the robotic program, to the near-term human program, and to the long-term exploration program?

And I know that there are a lot of people in the audience with a lot of good ideas about this, because I would hear them, as I was scarfing down my dessert at the dinners and lunches. And I think it would be good to get a dialogue going. How do we make this conference useful, rather than just all going home and having had a fun and interesting time?

QUESTION: Andy Presby. I'm student here at the school. I'm glad you asked . . . The sign on the wall there behind you says, why do we explore? And I think I've heard a lot of very inspiring and interesting stories over the last couple of days about why individual people—panelists and people in the room—have chosen to explore. And a lot of them are the same reasons that NASA has inspired me pretty much since I was born, since I can remember. But I think an important thing for you guys to realize is that the first thing that struck me is [that] not all of you explore for the same reasons. And when you're looking at NASA from the outside in—and I think some of the folks from Hollywood and the media have identified it correctly—the public is not as concerned about risk maybe as the explorers are. The public seems to have sort of understood that you guys accept the risk and you do it because you love it, for whatever reasons.

What the public is worried about is, why are we going? And why should I pay for what you love to do? Why is it helpful to me to pay for what you love to do? And I think that if you guys walk out of here with anything, perhaps a useful thing would be an internal dialogue amongst yourself culminating in an intensified outreach program to explain to the public, in terms that they can understand clearly, why they should pay for what we all in this room, I think, would agree is one of the most important things our government does for us in this country.

MICHAEL FOALE: I think we heard very eloquent expressions over the last three days as to why we explore. I'm actually more worried about the public not perceiving when it's dangerous. I don't believe people expected the *Columbia* accident. Astronauts do expect the *Columbia* accident. And I think there are misconceptions out there. Someone referred to it. It's the repetition of anything that makes us numb to the risks. And because we've seen Space Shuttles launch and land successfully a number of times, it was a surprise.

The risk has not been well communicated. What *SpaceShipOne* did today was extraordinary. And you saw, if you were watching, how something very unexpected happened during the ascent. The Space Shuttle goes—there's 25 times more energy in that whole business. I mean it's 100 tons, is it? Take the speed, divide by 25, and square it, you're going to get the answer.

JOHN GRUNSFELD: It's 25 squared over 3 squared. [laughter] We're two physicists. We'll have this in a moment.

MICHAEL FOALE: So if it's Mach 3, and we go Mach 25 in a Space Shuttle, divide 3 by 25 and you get about 8. And then you square it, and it's 64. But it's huge. The difference is that their heating on entry is just going above boiling—if that. It's not anywhere near risking a metal hull. If it's a composite hull, it's going to start risking it pretty soon.

A space vehicle gets up to 2,000–3,000 degrees Fahrenheit. So these issues are engineering issues, they're mundane, they're arcane to the public, who don't really care to hear the details. But the final answer is that it's dangerous. It's risky if any of these things fail.

John brought up an interesting comment about the risk of this launch today, which I would like to get to, to tell you that there is risk here even in *SpaceShipOne*. He talked about the lack of a pressure suit. He talked about the need for closing the hatch and living only in shirtsleeves there. We don't do much different on the Space Shuttle. We have pressure suits, and we have parachutes. They didn't do the *Columbia* crew any good. I don't know they would have done the *Challenger* crew any good.

So the situation really isn't so different. And, yet, John pointed to the risk this morning for Mike Melvill as he did that climb. The risk is still there for every Shuttle astronaut that will be flying on the Space Shuttle henceforth.

JOHN GRUNSFELD: I was trying to use the seatbelt analogy. It doesn't guarantee it. But it reduces the risk. Good point.

CHRIS MCKAY: Why does the public think then, that NASA is going to make it risk free? There is the perception that if we were disciplined, if we followed the *Columbia* accident report rules, and if we had a culture of safety, we would be risk free. Somehow the message that you guys are saying, which is that it is inherently risky, people are going to die, crashes are going to occur, is not being conveyed by NASA. We're not getting across the message that you're articulating.

And that's what I'm saying is, how do we turn this conference into something useful? Well, maybe we need to start figuring out how to get that message across, and stop giving the impression that we can make perfect systems.

QUESTION: Tom Krause, BST. We're involved in assisting with the culture change effort at NASA. It seems to me that the issue is not so much that the public doesn't recognize the risk, but rather that the public finds unacceptable the possibility that something could have gone wrong organizationally that led to the accident. So, when the investigation finds that errors and mistakes were made that could have been prevented, then it seems to me the public says, something about this just isn't right.

CHRIS MCKAY: Can I react to that for just a bit? If you take just about any accident and trace it back, you can find a step or a place where it could have been prevented. That's just the nature of these complex systems. And I don't think that you'll ever be able to come up with an institution, a large group or even a small group, where your accidents due to human factors or human error are gone. I think that's unrealistic. And maybe Scott, who is on the Board, we might put him on the spot here, since I think he works at the same Center I do, could comment on it.

SCOTT HUBBARD: Yeah, let me see if I can parse a little bit from where you're coming from, and what he's saying. It took a long time to be sure we had the physical part [about the *Columbia* accident], in the end we got that with no equivocation. Everybody absolutely knew that. The organizational part took a lot longer, or took a different approach, and was in many ways more complex to understand. And, I think, having people come in and talk to us, having members of the community as well as experts in behavior and complex systems and human factors talk to us, the distinction was that we had, perhaps, led the public to believe that we had done everything we could reasonably do. And, in fact, as we peel the onion on the accident, we found that there were cases where, because of repetition of something that started off as an inflight anomaly and became a turn around issue, because of other situations where people had fallen into poor habits of engineering analysis and so forth, we really did make some human errors that, with a different type of approach to it possibly, could be addressed. And, in fact, that's the result of the culture change.

So, now what we have to do, I think, is to tell the public that there is a level of risk. That we are doing everything we can to mitigate that risk, but it is not going to go away past a certain point, there aren't perfect systems. We are going to address the culture issues as much as possible, but there is going to be an irreducible residue in there that you're going to have to deal with. I think the danger is that, with the talented people in this room, and the Astronaut Corps in particular, you make it look so easy. All the thousands of people that support, with all the things that are done, the impression comes across—whether it's in the robotic program, with the perfect landings of Spirit and Opportunity, or whether it's with the Shuttle program with, by all accounts, a perfect takeoff and landing—that we've got it down.

The fact is that anybody who has participated in a launch, particularly if you've been in the position of being the last person to say go, and you hear in the background, through your earphones, all the thousands of things that have to be right, all the systems that have to be polled, you know that there is an irreducible risk of something catastrophic happening.

We do not tell the public that story. I think if the public just had the earphones on of the guy in the polling chain as you're getting ready to launch

225

and was aware of all the people at all of the systems and all of the things that have to happen, they would immediately realize, gee, what we have to go through to make this happen, it's truly extraordinary. So, I think that that's part of what we need to communicate, and part of what this business is all about.

JOHN GRUNSFELD: Let me put John Gans on the spot. He's the director of the National Outdoor Leadership School. They have thousands of students every year who go out into the wilderness, go out into risky situations. I imagine occasionally a parent will call and say, my son or daughter is going to go out and do this rock climbing, is it safe? And how do you communicate to them the risk element as an institutional risk manager?

JOHN GANS: Well, first off, we try and be as clear as possible that we can't guarantee anyone's safety, and we're up front about that. I think every time I get on United Airlines and I hear, you know, "Safety is our number one priority," it runs through my mind that, no, getting us there is the number one priority. Safety may be number two. But say safety is number one, we wouldn't take off.

JOHN GRUNSFELD: It's clear that profitability is not number one.

JOHN GANS: So we try and be as clear as possible. And you mentioned the parent-child thing. I'm going to switch the question some, because Dr. Sylvia Earle talked about the role of education as it relates to exploration and risk management and getting people outside and other things. And I think the interesting thing that I have been thinking about in this conference is that my daughter, this summer, started climbing in a more aggressive way. Safely, but in a more aggressive way. My daughter was 10 this summer, and she wanted to get ready to climb Devil's Tower with me this fall.

I adore my daughter, you can probably tell. And it really hit me that, suddenly, I'm on the other end of this, and I'm hesitant about what she was going to do. Now, climbing has been one of my passions in life. That's where I've felt most alive. It's where I've had some of my best relationships with people around me, with the world around me, and the environment around me. And, suddenly, was it okay for my daughter to do it when it moved beyond the walk-up situation into something that was more serious? And I came to terms with it. We are going up to Devil's Tower in October.

But there is something about generational passing as it relates to risk management. And we certainly run into it with parents making decisions for their children. It certainly is tied into the educational issue. But it's something that goes to each individual family, and it's something that I've thought a lot about over the last few days. It's something that goes to the space program, the generational difference between the people that grew up with Apollo, the generational difference now. Look at the number of parents now that won't let kids go off and ride a bike alone, wander out of their neighborhood alone, whatever else is the case.

I realize I'm broadening the issue far beyond NASA here, but it goes so far beyond what we're talking about here, and, somehow, I think there is a role to play for our society in making the parent-child relationship understand risk better. And there is a role there for our schools. So, now that I blew up your question totally into something else, I'll pass off the mike and not go on further. But the long and short of it is, we try and be very clear with parents that we can't guarantee any safety out there, but we manage it very well, and then we convey the benefits. And we know the benefits right down the list, and we rock at all those benefits, and are clear about them, much the same conversation that's gone on here.

JOHN GRUNSFELD: That's great, and it brings up another point, that you brought out, which is, if you think about the early part of the space program, prior to the first American going into space, rockets generally blew up. Most rockets blew up while we were trying this.

UNIDENTIFIED SPEAKER: They still do, John.

JOHN GRUNSFELD: They still do. Not most. Some. Let me just take this a little further, which is, when Michael and I were growing up, that's what we saw. We saw the struggles, there was no question that it was risky. And as we started flying more, and then we built a spaceship that looks like an airplane, it brought it into everybody's daily experience. Then, people who are growing up now, like my children, space is part of their culture. It's become the norm. And so people don't really notice the space program now until we don't have one. And I think that's an indication that it is part of our culture, and that the education can help. Go on, Mike.

MICHAEL FOALE: I just want to add that people in this room are probably aware that . . . I don't know of any rocket system that can launch 1,000 times and not have an accident. Most rocket systems launch 100 times and have an accident. So if that is the only way, if you're on the rocket on the 100th time, and you do a lot of trials, and you do the statistics, that's [it], you die on that rocket. So the way you get better than one in a hundred on any rocket system is to have a way of surviving that explosion that 100th time. And the Russians have done quite a good job with the Soyuz escape system, it's worked twice in all of their launches, hundreds of launches. Apollo was a good system never used. I think Gemini has an interesting case. It's a story as to why they didn't have an escape system quite like the Mercury before it.

But that is the way we get away from those—the fact today is that rockets do still blow up, and we can't do anything about it right now. We don't have a strange, wonderful, anti-gravity technology that will get us away from that.

JOHN GRUNSFELD: And one in a hundred is the best of the best. Most are not nearly that good.

QUESTION: Joe Fuller. I'm sitting here very anxious, because I don't think we're getting down to business. It's been a wonderful conference, you know, over the last two and a half days, and I think we've learned a lot. The problem is, how do we capture the knowledge that's been just flowing out here?

At some level there's a connection with the way we do business, and we need to search for that. So what I would suggest is that, the first thing, we capture the proceedings of this, and the second thing is, we form some kind of organization, you know, ad hoc or whatever, to pursue this information and make the connections that are so obviously there.

[In] some kind of way, the institutions have got to get involved in this. I think that, as someone said, every individual has a value calculation that they have to make, and they have to make that trade. We can't determine the perception of risks for the individuals. We can't determine the value for them. But what we have to do, and what we do in business is, the value proposition has got to be so large that the risk is acceptable.

So what I would suggest is that you've got to go farther than this. You can't stop here today. You've got to put some organization in place to carry this forward, and mine this knowledge for the value and the benefits that are obviously inherent in it.

I'm involved in risk management professionally. I haven't seen too many other people here that are. I did hear Mike Gernhardt talk about how he's using quantitative risk analysis. So I would volunteer to be a part of that group, to determine a strategy for extracting the knowledge and information so that it would be more useful and of value as we go forward and explore.

CHRIS MCKAY: I have a suggestion. I think that's a good suggestion, how do we connect to the institution of NASA, in particular, the results of this conference? It seems to me [that] to do that you need someone who is close to the Administrator. He clearly wants to get advice on this topic. Someone who is passionate with experience in this area. Somebody like the Chief Scientist, John Grunsfeld. I think we should add to his responsibilities this area. I think this would be a perfect opportunity. You've seen the conference. You were obviously one of the ones who put it together and organized it. I really think that the mantle falls on you to carry this forward within NASA as an institution, not just the near-term flight program, the return to flight. But also thinking long down the road.

Also, I think the robotic program is in need of a clear-headed assessment of risk. Now there the risks aren't to lives, but they are to resources. And I think that that program also needs a clear risk assessment. And I think the Office of the Chief Scientist right now is a good place to do it. So, all voting for John as the representative of this?

JOHN GRUNSFELD: Thank you, Chris, for your kind comment. But, seriously, I think we have Tom Krause here from BST working on our culture. This is something that the Office of the Chief Engineer, that my office, the Office of the Chief Scientist, Bill Readdy, Office of Space Operations, Space Operations Mission Directorate— this is a dialogue we have everyday. And we wanted to broaden that from NASA management to you folks, and, as I said up front, the start of a dialogue.

But the other point was capturing this and you were just captured. You were captured on videotape. We're going to convert that. We've been talking, prior to the conference, about how are we going to put all this together. A number of folks have been chronicling this individually, but we're going to do it institutionally as well. And I know, Keith, you've thought about that a little bit. Do you want to say anything? Let me put Keith on the spot, and then back to you.

KEITH COWING: I think we are quadruply redundant here. I am recording this on my iPod, so I can be listening to it as the transcripts arrive in my e-mail box in about an hour up at Ames. We hope to have this online in a very short period of time, just the raw verbatim transcripts, with the "ums" and the "ahs" and the spelling errors taken care of. I'd have to talk to Bob Jacobs and some of the Ames folks to get the specifics on what the follow-ons are, but there is talk of putting some of this on a DVD, of putting a more comprehensive document together. Steve Dick and I have talked about something more comprehensive, in terms of a history monograph. So the initial concept here, John saw the first e-mail that started this, was capture everything in as many ways as possible, so Joe, you're psychic, you knew what we were doing when we were first doing it.

QUESTION: Scott McGinnis: I'm a student here. What we do in the military, and I'm sure a lot of you are military, but if you assume this conference is like a six-month deployment, every time before you leave the ship you have to give your lessons learned. You get one line, everybody. Then the XO, I think, Dr. Grunsfeld, that's you in this case, forces everyone to read it prior to going on [to] their next deployment.

So that built a database, and as the XO you are required to make sure that they all sign and verify that they have done this, proving that they have read it. Then, when they make the mistake again, the responsibility then lies with the responsible individual, the person making the mistake. Therefore, you have a traceability and a responsibility for each individual action, and also, it shortens the amount of data; instead of having to watch our three days of deliberation, being able to shorten that and pull out the small pieces.

So that's part of the military structure, and you've got a little more discipline—I think we talked about the flogging and all that kind of stuff. [Laughter] We have a little more coercive nature in the military to be able to do that. But I'm sure NASA can muster that up. And second, you're talking about the [pressure] suit of *SpaceShipOne*, the risk that they're taking. And I think it all goes back to the benefit that we haven't discussed, we've tapped around it. Dr. Spudis brought up the three reasons why we explore.

I think the fourth, and Magellan showed it with his cloves, is money. And *SpaceShipOne* is doing it, one, to explore. But come on, we've got a $4.5 billion market in the tourist industry. And they are exploring not because they want to prove science or prove humanity. We've proved we can do it with the money. But can we make it profitable? And I think if you saw the big "Virgin" on the side, and you saw the big Sprite advertisement going on, and the M&Ms floating around, I think we have found one of the keys to space exploration, and that is the good old American greenback.

I think NASA also needs to find a way to maybe encourage that, like we did in the early '20s with the prizes. I think we've tapped around cost as a benefit, or money as a benefit. I know NASA can't get benefit monetarily that way. But it's definitely a point we haven't brought up.

JOHN GRUNSFELD: No, we'll be offering prizes. We've got Congressional authority to do that for similar challenges. But it's clear that for Rutan, this is about his passion, about pushing new envelopes. The X prize is $10 million [he blew off] quite a bit more than that. And I think it's great that he's been able to leverage the commercial sponsorship there to help offset his cost of developing this Because it is opening a new frontier.

Eventually, folks who want to actually sell services will have to start incorporating more of the safety rigor. You probably wouldn't go out on a cruise ship today if you knew that one out of every four or five times you weren't going to come back. So that's, again, that comes down to the profitability. And what Mike Gernhardt said is, you have to have a successful dive operation to have a commercial operation be successful. And so, safety is a critical part of that greenback.

QUESTION: David Liskowsky from NASA Headquarters. I'd like to perhaps comment on some of the discussion that's been going on. We're at a point in time at the Agency where we've just gone through a large transformation to hopefully meet the exploration vision. I think we're all behind that, and that's what we'll be going forward with.

Maybe we can take this opportunity at this time to use these changes that are going on in the Agency to change our message. Change our message to the decision-makers, mainly Congress and the public, about what the nature of this business is, that it is risky business. Everyone talks about that, that NASA has been a victim of its own success.

But maybe it's time that, as we go forward with this new exploration vision, and this is something that can be done through John as Chief Scientist, we have the PAO [Public Affairs Office] folks who shape the NASA message let folks know, truly, what the nature of the business is, and to let them know that, as we go forward with this new vision, it is going to be risky. And without abdicating our responsibilities to meet the requirements of the CAIB [Columbia Accident Investigation Board] report, there is going to be that element of risk. And it's part of us shaping the Agency's message and how we convey it to the public.

Maybe this symposium can be the first step in trying to do that, in shaping what that new message should be for the exploration vision, and making, perhaps, a little more realistic vision than the Agency has had in the past.

JOHN GRUNSFELD: Good comment. Well, last comment, and we'll go to "Moose" Cobb—Robert Cobb—he's been dying to say something. Then we'll stay around for comments afterwards. In the packet, we gave everybody a pad of paper and a pen. And so, before you're allowed to leave, you'll have to write down at least one lesson learned, and provide that. You don't have to have a name on there, but if you have any ideas, sketch that out and put it in the bin before you take your STS-114 pin, which I really want you all to have.

ROBERT COBB: I'm the NASA Inspector General. I've been with the Agency for two and a half years, with no exploration or science background coming into the Agency. It's my perception that NASA works hard to dispel the notion that what it does is inherently risky, and the reason it does that is because there is a fear that the public won't fund it if NASA tells the truth about the risk.

That's something that I think that this conference goes a long way towards— I think people recognize that the public is willing to accept risk. And that the idea is, the object is, that it's important for NASA to have a transparency into the risks that it is accepting and to allow the public to share in understanding of those risks.

QUESTION: I'm Sandra Cauffman. I'm from Goddard. I think we're missing some basic thing here. The question is, why we explore, and we are not really answering that question. We're talking about the risks and, yes, that is very important, but the people out there need to understand why it is that we're doing what we're doing and what they are getting in return. They like to understand why we are risking the people, but what are they getting back?

In the DOD world they understand why we are risking our soldiers and why we are sending people to war and whatever, but in the NASA world they do not understand why are we sending astronauts. And they see pretty pictures of the stars and stuff, but what is it that they are getting back in return as taxpayers? And we need to really send a clear message to them. And it's not PAO [Public Affairs Office] stuff. It depends on each and every one of us to do that.

Just a little story. I was in National Night Out in my neighborhood a couple of years ago and I was talking to my neighbor, a nice little old lady. And I am the Deputy Project Manager for the GOES-R Satellite. And she was asking me what I did for a living and I told her about the weather satellites and all this and all that. And she just looked at me with this puzzled look on her face and she said, "Why do we need weather satellites when we have the Weather Channel?" You know, that's what we have to deal with, the perceptions out there. Yes, the risks are there, but they need to understand, okay, we are risking, but what are we getting back? So, I just wanted to say that.

JOHN GRUNSFELD: That's a great comment. Natalie, why don't you take that?

NATHALIE CABROL: Actually, I would like to add on that comment because this is probably translating better, what these guys were saying yesterday. What is the gold? Not the goal, but the gold, you know? Five hundred years ago, Magellan leaves, and he brings back cloves and he brings back riches. What are the riches that we can show to the people today? And there are many. And we are good at it at NASA, but we are not good at telling people. You know, from the Moonwalks people today are going to ski better. They have good Moon boots, medication, things that we do in space better the health of people, the expeditions in the sky, in the sea, or on the land are bringing [generating important discoveries]. We have that, but we are not translating enough to the public. And I think this is where we need an effort.

JOHN GRUNSFELD: Steve?

STEVE DICK: Following up on that—I'm Steve Dick, the NASA historian. On Friday we're launching a series of essays on the nasa.gov Web site called "Why We Explore." And I think this will address some of the questions just raised. And it's not Public Affairs, it's historically nuanced and historically based. (And, by the way, this is the 46th anniversary of NASA, on Friday, October 1st.) The first essay will deal with why we explore in the sense that exploration is necessary for a creative society. And I'll talk about Ming China, which was mentioned by Jack Stuster the other day. That's on the NASA Web site at *http://www.nasa.gov/missions/solarsystem/explore_main.html*. And it'll be a once-a-month, "Why We Explore" series, a different essay each month.

NATHALIE CABROL: I will wrap up quickly. But, you know, why do we explore? I think within us it's just because we think that somewhere on the other side of the hills, as you were putting it, it must be better or something is better than what we have now. Otherwise, we wouldn't be doing it. And it's true that maybe the other side of the hill has nothing particular, but what we learn along the way is bringing a lot of good to society, et cetera. So we need to emphasize this really, really hard.

JOHN GRUNSFELD: And again, it comes down to both personal and institutional, as well as national. In the President's vision [for space exploration] he said, "The purpose of this is to advance U.S. scientific, economic, and security interests." And it's through a broad range of things. As you say, along the journey you learn a lot of things that improve our life here on planet Earth.

But it's also the higher purpose. You know, we're trying to understand where we came from. Why is there a universe? And in the process of very basic research like that is where we learn the really valuable things—like quantum mechanics that leads to lasers—that it would be a long time before you'd do that with just subsistence farming. So, these types of things are very important.

I also have something that often ends up resulting in controversial discussions, but I have a statement that I think is true. I can't prove it, but it's "Single-planet species don't survive."

QUESTION: Dave Leckrone, Hubble Space Telescope and NASA Engineering and Safety Center. I guess we're all ganging up on you because several of us must have made the same comment to you. So, I want to start out by thanking you for stimulating this conference, which has been absolutely fascinating. What fascinated me most in hearing all the speakers and the discussion and seeing the film last night about Ernest Shackleton and the *Endurance* expedition was this business of what compels us to explore and take these risks in the first place, instead of just adopting the fetal position in our lives.

And I have my own ideas. I actually wrote it all down and I'm going to exchange this for my pin later. But it sort of goes to what was said just a moment ago, and I think Scott Hubbard mentioned this on the first day. We explore because we have no choice. It's an evolutionary imperative. Our species became what it became because it explored. What was over the next hill was either a threat or a source of sustenance. And if there wasn't anything there, then you had to go to the next hill yet to check that one out. And I think this is built into our DNA.

Poor Ernest Shackleton was so obsessed with exploring he couldn't even really articulate why he kept going back to the Antarctic. He just had to do it. And I think at least some of us, if not all of us, within the species have it built into our DNA. And I think corollaries to this are all having to do with survival—acquisition of knowledge, commerce, education, creating a national identity, finding not only individual self-fulfillment but group fulfillment. And I think every one of those relates, going way back perhaps, to our need to survive as a species. And maybe we can't survive as a one-planet species.

JOHN GRUNSFELD: I agree with that absolutely. You know, we try and raise it to a higher plane but, ultimately, it is, I believe, hardwired into us to do this. But, as well, our evolution has taken us to be a species which is a thinking species, sometimes rational species. And, so, it's also provided us the ability to question what we do. And that's where this becomes a little bit messy, because we say, "Well, is it worth the risk?" And that comes back to where we are.

And if anybody doubts that we have a survival imperative to explore, just look at the situation we're in with science, technology, engineering, and mathematics in this country and where that may lead to eventually—because technology is the key to economic prosperity, which is the key to security, which is the key to freedom. And I believe that exploration is linked to our ability to stimulate people to, directly and indirectly, get a good education and make use of that productively.

UNIDENTIFIED SPEAKER: I'm just going to build on some of these other things that people have been commenting on. And, in particular, I want to play devil's advocate to some of the spinoff comments that have been made.

I agree that this is very important, and some of the discoveries have been fantastic. But really—and this builds on your comment earlier—I think that there's one question that NASA needs to be accountable to, or one big question, and that's quite simply, are we pushing the frontier? Are we pushing the frontiers of science, technology, and exploration in a way that no one else can—no individual, no company, no university—in a way that only NASA can? And that's the thing that we constantly have to be asking ourselves. And I think this conference is part of getting at that issue.

JIM GARVIN: Well, thanks, John. I think there's one comment notwithstanding the spinoffs and everything. I mean, we can all play the game as, Dave, you said so well about, this is an investment choice. It's part of our DNA. But I think it also bears witness to trying to generate metrics and look at what the impacts have been. And we do that perhaps ineffectively, as you've said, Nathalie. But they are not transparent, they are major. And if you ask some of the technology leaders, exploration has begotten these catalytic effects.

So without it, the question that you raised, John, is the one I think this group needs to raise. How fast would we have progressed in different areas? I mean, maybe Darwinian progression—you know, seeking optimization whenever we can—is not the game afoot, and natural selection in technology doesn't work. I don't know. That's a great thing to debate. You know, maybe Steve Dick's group in history can study that.

But I'm still struck by questions that when we ask people in other sectors of society—IT [information technology] being a good example—in remote sensing of this planet, the benefits, while maybe not tangible in terms of dollars in your pocket, are there. We would not have microcomputing with fault tolerance, ever. There would have been no imperative, except perhaps a very narrowly-defined security interest area—which is important, of course—without this exploration imperative. And we demonstrated that.

So I think we need to do better at defining those metrics. I mean, yes, the textbook metric, I think, is an important one that most people seem to forget. I like to think that all the textbooks have been rewritten in the last 20 years in many of the areas of astronomy, physics, planetary science, and even this place of our own planet.

But anyway, I think that's the amplifier on technology progress in areas that aren't the ones that have instant economic gain. That's what we should be doing, and that follows on what you said so well. That's NASA's unique role as a government agency. Otherwise, it would be private. Thanks.

QUESTION: I'm Becky Ramsey, NASA Headquarters. Recently we had someone do a study for us. And while it was a very interesting study, I won't go into the whole thing. But one of the stats that struck me is that a majority of the people we talked to said that they like NASA. They don't have a clue what we're doing, but they like us. And I think we cannot lose sight of the fact that we're not the only ones who want to go. It's not confined to the people in this room or the people who attended this conference.

I walked over to the little lobby bar last night. I was sitting there watching the baseball game, and I got into a conversation with the bartender and some of the servers. They said, "Are you with the NASA group?" "Yeah." "That's so cool!" You know, they don't know what we do, but they like us! And we have to build on that personal connection. We are their representatives. Until Burt Rutan starts charging five bucks for a trip into space, most of the people out there are not going to get to go. We have a responsibility to be their representatives and to do what they can't do yet. I mean, we talk about the spinoffs. They don't really care about the spinoffs. Yes, they're important. Yes, the benefits that we [generate] make everybody's lives better. But they don't know about that, you know? We tell them, but they don't read our cool little magazine. They don't know the weather satellites from the Weather Channel. They don't care that much about that. They like it because it's cool, because they want to go. And I think we can't lose sight of the fact that that's why exploration is important to everyone else.

JOHN GRUNSFELD: I absolutely agree. In fact, in other studies we've found that the NASA logo—the meatball—is likely the number one brand recognition. There may be a couple others that are close. The other thing we found out is that, when we were working on our renewed vision of discovery, we found out that most people assumed we were already doing all these things. You know, when we'd say, "Well, what do you think about having a renewed trip to go beyond low Earth orbit to the Moon and Mars?" folks would say, "Well, isn't that what you're doing?" And we'd say, "Yes, that's what we're doing!" And we have to communicate that a lot better.

MEL AVERNER: That's not true. We're not going to Mars and Moon. We are attempting to do that, but it's not our mission yet. And if we say, "Yes, we're going," people will go away saying, "Great! Great! You're going!" Okay, you got my drift.

JOHN GRUNSFELD: I wish Steve Squyres were here right now. I think he would argue with you. He has two of his children on Mars right now. I don't know. Jim, do you want to comment just a little bit about our program, what some of the next steps are that are already in place?

JIM GARVIN: Yeah. Well, I think maybe Steve would do it better but if you don't think we're exploring now, maybe we *don't* communicate that well. But I think—two rovers 270 days on another world wandering at 300 percent beyond expected lifetime is a new demonstration of that. Cassini alone is exploring at the highest order.

MEL AVERNER: I'd like to respond.

JIM GARVIN: But, let me finish. I mean, I can go on and on with the legacy of how we explore. It's just that, right now, a lot of people, perhaps in the public sector—and I can't speak for them because I'm a geek and work for NASA—but when I talk to them at hockey games and things where they don't always care what we do, they're stunned by what we're doing and how we're exploring. And how we've learned to go from people on the surface of the Moon as our agents of exploration, being our representatives, to machines being those agents. And we're doing that so many different ways. We're so diversified. In fact, if you ask corporate America and many of my colleagues there, they're stunned. "You're doing all that, with that portfolio? You're nuts!"

MEL AVERNER: I'd like to get back to the bar last night. Becky, was that your name, doing what I would have done—drinking at the bar? Suppose you were to go back to the bar and talk to those people and say, "Well, we are exploring. We have two robots on Mars doing terrific scientific things." Would they say, "Wow, that is great, but when are we going?"

235

JOHN GRUNSFELD: Absolutely, I agree. But just to give you the counterargument—and I don't know what the current number is, but there have been 13 billion hits on the NASA Web site of which three-quarters . . .

UNIDENTIFIED SPEAKER: That's a false number. It's not 13 billion people.

JOHN GRUNSFELD: No, no. I didn't say it was 13 billion people.

UNIDENTIFIED SPEAKER: I know, but that's the impression that it leaves.

JIM GARVIN: But there are well over 100 million unique IP addresses, maybe 250 million total. It's all around the world, predominantly the U.S., but all around the world. And, you know, you could argue about the numbers, but it is so much greater than any other Web site that it's phenomenal. There is interest there, and there's interest specifically because, I believe, that what we've done is we've put two human eyeballs on the surface of Mars. So people see what the rovers see and they think, "This is kind of what I would see when I get to go." Or, "When we send people, this is what they will see." And we want to do that.

QUESTION: I'm Nancy Ann Budden, Naval Postgraduate School and Lunar Planetary Institute in Houston. I want to build on some comments that were made by Joe Fuller and others about getting the word out and on some communications issues that Jim brought up. I joined Johnson Space Center's Exploration Office in '88 and I worked with a lot of you, Chris and Dale, on human exploration issues, and this was about the time that Bush '41 came out with his announcement that we were going back to the Moon and on to Mars.

One of the things that we neglected to do over the next 12 years, really, was put into place a communications plan. We all had great ideas. We had a lot of meetings. And now we have another opportunity with Bush '43 coming out with a much more reasonable, cost-rational plan and vision. And one thing I think we really need to do is put together a communications strategic plan, like a mission, and have a schedule and a budget and have somebody own that. Whether it's PAO [Public Affairs Office] through NASA Headquarters or whether it's an industry/NASA/university team. But we need to have a plan for that, that actually has someone own it, someone that's going to pay for it, and understand who are the advocates that we need to build. Obviously, there are communities we need to get to within NASA, of course. We need to get to the [Capitol] Hill. But we need to do it in an integrated, planned way with someone thinking about, okay—who are the first people we need to get to, and when and why, and how do we integrate this message? I nominate Keith Cowing to put together the message [laughter]. And, John, I think everyone would love for you to run the communications strategy idea since you're getting asked to do a lot of other things this morning and since you have a lot of spare time!

Anyway, I would like to see someone own that and put together a message that people agree with and actually stand behind, and make sure that it is consistent with our Commander in Chief's vision of the future for space exploration.

JOHN GRUNSFELD: That is absolutely a great comment. We've received that comment quite a lot, so we've actually heard that message and we've acted on it. Part of the transformation was to create a communications group, and we've linked the legislative and the public affairs and our external relations into one team so that we can help craft it. We were at the bar as well last night, talking about a budget, specifically, or an increased budget, line items, and management for public affairs as well. That's crucial, that we have to treat that as something that's very high-priority. But in the transformation, we've combined all of those for exactly the reason that you mentioned. Thank you.

QUESTION: David Gast. I'm the other student here for the school. The thing that I think everyone here is touching on, and building on some of the things that have just been said, is it is about communicating to the public. I think everyone in this room and most of the people watching NASA TV already know, kind of, the reasons that we want to go out there, what we hope to accomplish, where we hope to go, and understand the risks that are inherent to doing that. With this communication message, what we have to do is say to everyone else, the people that aren't in this room and aren't watching NASA TV, "This is where we want to go and this is why we want to go there. And, you know what? It's dangerous. Very likely, things are going to crash. Maybe people are going to die. But the people that are putting themselves on the line for that understand that and accept those risks for themselves and believe that the goal of what we're trying to accomplish is worth that risk." So, I think it's all these things.

We have to communicate the risk, yes, coupled with why we think the risks are worth taking. We can't just say, "We're going to do these great things, we're going to go to Mars, go to the Moon, and it will all be safe and happy and fun." Neither can we say, "It's dangerous to travel through space." We have to say all these things at the same time.

We talk about [that] the American people won't accept that something went wrong that we could have avoided. There's always one more thing we could have avoided had someone happened to think of it, had someone happened to see it. And I think they're willing to accept that if we're doing the best we can with what we have, there are always dangers there. And they're willing to accept that, again, if we communicate that to them in advance. Like I said, the people here all understand that. We need to take what we've talked about here and present that to American people.

JOHN GRUNSFELD: I think it's T. K. Mattingly who told us, "Success always has failure as its predecessor." He was more eloquent.

QUESTION: Keith Cowing. Thank you, Nancy, for the nomination. When you hear what I have to say, you may withdraw it. To the point of Web traffic—and you're right, I do Web sites for a living—citing Web numbers is so 1997, so Pathfinder. [laughter] Google does that traffic before lunch on Sunday. It's great to hear these numbers, but I could go write something in my room right now.

Drudge Report would pick it up and have a million hits by tonight. Big deal. The Web hit numbers are important. A lot of people are looking at NASA's Web sites. But we need to move on to other metrics. When a nine-year-old girl raises her hand at a Presidential visit and asks about space—things like that—then you know. When the late-night shows make different jokes about space—Jim Garvin has done yeoman's duty, going on *Letterman* and so forth. When you start to see this consciousness of space percolating up in other places . . . These numbers can be very deceiving. Anybody can generate hits. You've just got to look for other metrics. You've got to have a new metric every month. Just some advice from somebody who does this for a living.

UNIDENTIFIED SPEAKER: This sort of follows your point, Keith. But when the NASA crews come into the small town of Lander, Wyoming, to go on [National Outdoor Leadership School] courses, they often stay after and talk to the kids attending courses. And when those kids leave the room, they're changed. And following on your point, I believe, it's not about communicating to the public, it's about changing the public.

KEITH COWING: As the Administrator of NASA loves to say—it is this Jesuit thing he has—"one conversion at a time." It works. [laughter] It's self-propagating if you do it right.

JOHN GRUNSFELD: I should say that every time an astronaut leaves the school, they're changed as well.

QUESTION: Bill Clancy at NASA Ames. One concept that we haven't talked a lot about here that I found very useful as it relates to the public, and also inside, is the word sustainability. To me that's the most important word, I think, that's in our current vision. And I found it very useful to the shift from thinking about particular missions to the program. So, rather than just talking about mission risk, we have program risk. And we're talking about building competence and the ability to go places and so on.

I first understood this, I think, with Mars Polar Lander, where we didn't have the telemetry that we needed to give us the information for building the redesign that we needed. I think your example this morning is a beautiful example as well, of the investment that one can make to build tools that will give us a competence that we know we want to have [as] part of our tool kit. So, I think when we're articulating to ourselves what's our priority and our objective, it's the clear objective, maybe dates, and the sense of challenge. But it's all about sustainability, and we make decisions because we need to be here tomorrow. We're not going to climb Everest today, because just getting to Everest today is not our goal. We want to be able to climb again tomorrow.

JOHN GRUNSFELD: Anyone else? David.

DAVID HALPERN: Thank you, John. And one of the things we've learned—some of us knew before, but some others learned—that 96 percent of space needs to be

238

explored and 96 percent of the ocean needs to be explored. One [space] has zero pressure and one [the ocean] has a very large pressure on the bottom. And then the question comes that the ocean definitely is a place to explore, for two reasons. One is [for] the creation of new knowledge, which is the same as what you're talking about for outer space. But inner space also has a well-recognized aspect of creation of wealth. I mean, a number of [beneficial] activities have always gone on in the ocean—and I don't mean just transportation, but subsurface as well—and new ones are coming along, like genomics, oceanography, things like that.

So, then comes the question. In the new, transformed NASA, the challenge, now, would be to make use of the fact that oceans—or inner space—require the same type of dedication and the same type of methodologies as are being used in exploration of outer space, and it's something that the new NASA might want to consider. And it's actually well-poised for that because all of the science now is in the Science Mission Directorate. Rather than in two different stovepipes, it's all in one. It's a comment, not a question.

JIM GARVIN: I'm really grateful for you for saying that because my new job at NASA, with the many hats, is, in fact, to try to integrate the inner and outer space exploration in this new vision. So I'm looking, as is Ghassem Asrar [NASA's Science Deputy Associate Administrator] and John, we're all looking for the connections. Because I think the point with a vision, with an objective, with some of these good points about program-thinking, which we've had in EOS for Earth science, we've had in the Mars program, we hope to have throughout our program—the Shuttle program—is an aspect of risk that I think is the one that right now strangleholds a lot of us. And that is risk of our own interpersonal management structures to get the job done.

And that, perhaps, is the genesis of the transformation, to get around some of those things. But, you know, when organizations grow old they become well-rooted in certain directions. And breaking roots, it's like taking a root off a redwood out there. I mean, it's going to stay three hundred feet tall, so you don't want to have it fall over. You want to have it move. And other than slime molds, most large plants don't move.

But I think that's the challenge. The ocean is an exploration frontier that will teach us about high-pressure environments and knowledge and all that, and some shared technologies could be trialed there in the name of science and exploration to good end. And, you know, it's rather ironic to me that a large fraction of the ocean exists at 100-bar pressure, which is the average surface pressure of the planet Venus. And, you know, lots of living stuff there. Interesting to think about.

JOHN GRUNSFELD: Thanks, Jim. We'll take one more.

QUESTION: George Tahue from NASA Headquarters. Listening to some of the comments here, an analogy is coming to my mind. If you're familiar with the paleontologist Stephen Jay Gould and his description of evolution as punctuated evolution, I think NASA is, as a government agency, going through an evolution, and we will continue to do so. Where we're going with this is going to take a very long time, but there are certain points where there will be punctuations that make great changes in very short amounts of time. And I think Apollo, that era, was one of those points. We may go through slower periods of time where we go through those changes. But here we're at another point where we may be at another one of those punctuations. And this new transformation that we're looking at isn't just rearranging the deck chairs. And it's something that we have to take internally and not just focus only on, why didn't the public understand what we're doing and how can we make them understand? It's something that we have to do over this long period of time, even internally.

When we had our transformation and the Office of Earth Science and the Office of Space Science came together, I was listening to some of the Earth science guys and saying, "Wow, you do that? That's cool!" Same reaction [as] at the bar. So, I'd like to charge all of us to try to take a lot of this internally and focus on those goals.

Another key thing we've heard here is to focus on the target. Stay on target. Protect and understand our planet. Search for life. Understand the limits of it, and recognize that humans and robots are the tools to do those goals. It's not just, "Get us there." It's not just, "Get the robots there." Focus on those [larger] goals. We'll have these punctuated evolutions where we have a grand target that we're looking for. And, in between, we'll have this balance that we keep going forward in trying to get that message to the public to understand that we, as an agency, have a role as a public function in our society. So, those are my thoughts.

QUESTION: I'm John Gaff from the Glenn Research Center. I think the Agency, while it does wonderful things—and I've been in it a long time, is not recognized by our society as critical to the survival of society. Nobody questions why you've got the State Department, nobody questions why you've got the Treasury, and nobody questions why you've got the Defense Department, or Agriculture, even. But for some reason, we have been unable, in my opinion, to transfer the knowledge that we are able to acquire for the future to being something critical for the survival of the economic success of the Nation. And for the long-term viability of the Nation.

Somehow, we need to start some mechanism—and maybe it's in the education programs, these outreach things—where we get more institutionalized as a recognized, long-term investment. Until that happens, we're always going to be at the margin, we're going to be at less than half a percent of the budget, and we're not going to be able to compete for the other critical needs of, "What's in it for me?" with the society. Thank you.

JOHN GRUNSFELD: I think that's a very good comment. I would like to point out that we're in relatively tough economic times right now, yet NASA is the only agency that's basically gotten an increase in its budget.

UNIDENTIFIED SPEAKER: Did it get one?

JOHN GRUNSFELD: Well, in the request, in the request. And even in the appropriations meetings, we've fared better than virtually all discretionary agencies. I think the issue is: We're still a discretionary agency.

UNIDENTIFIED SPEAKER: I'd like to kind of second that and say two things. You guys are in a really tough position, almost a harder position than you were in the Apollo era, because Apollo was something we all thought we needed to do. You guys are in the very, very hard position of deciding what we should do. We don't necessarily need to do anything, it doesn't look like. There's not an immediate and obvious need. But you guys can do lots of different things. I'd like to tack onto Dr. Halpern and Dr. Garvin's comments. I think you guys recognized this, but I'm not sure that the rest of the population does—one of the amazing things about the way NASA is exploring the new frontiers in space, and the way that ocean science explores our frontiers here on Earth, is that for the first time, I think, in human history, you've got the conservationist, the naturalist, the scientist, and the greedy capitalist wound up, in many cases, in one mind, in one human being. And you've got an organization that's already looking to protect resources that we can't even exploit yet or use yet.

I mean, does that seem strange to anybody else? That's new, folks! I mean, I think even more so than technologies, you guys can share lessons learned and organizational experiences based on how do we commercialize this thing, and how do we get benefit out of it as a people without destroying it for ourselves and our posterity? And perhaps that could be part of your public outreach program, because, for instance, look at the market for the Toyota Prius cars. It's huge! They're back-ordered, I don't know how long. Eight months back-ordered on the cars! You know? The public gives a darn about that kind of thing and you guys do it [balance benefits and conservation] every day. It's innate. It's part of your nature. That's important.

UNIDENTIFIED SPEAKER: Coupled with that and, again, talking about expanding the vision and explaining the risk, is that the vision we want to put out there is not just [that] we want to go back to the Moon and learn how to go to Mars. But I think it's a bigger vision than that. It's partially this and partially the thing that he's talking about. It's [that] we need to present both, this is the next step on which we are currently embarking, but also, this is a vision for the future that we hope to achieve by taking these steps. And that vision doesn't have to be perfect. It doesn't have to be exactly what we're going to arrive at. But it has to be a goal beyond just, you know, as great as the goal was to put a man on the Moon and bring him back to Earth. Why? Now why are we doing that?

And we've talked a lot about that, but I think that needs to be part of what would go out to the public, and what NASA thinks about internally, and each of us thinks about internally, in ourselves, as what is our long-term pictured goal that all these things are steps toward? And that goes for exploring the seas as well. You know, all these explorations are not just, I want to go to the bottom of the Marianas Trench. It's, I want to find out more about the Earth. I want to discover more about us as a species. I want to maybe discover things that will save our species or our Nation or whatever at some point in the future. So that needs to be a part under consideration as well.

JOHN GRUNSFELD: Okay, well, I think we're up to the end here. I just want to give all of you a big "thank you very much" for participating in this. I know I've learned a lot. I think we've all had a lot of good dialogue. I got a few too many action items, but they're very important ones and we will take that forward, back to NASA, and for those of us here from NASA, I hope you take that all out. I really want to encourage you again, though, as you leave here, to regard this as the start of a dialogue. There's no question that this is one we'll talk about sustaining. I think this dialogue will be sustained probably for all of human history as we push our frontiers, as we move out.

I'd like to bring Scott Hubbard, the Director of the Ames Research Center, to give us some closing comments.

APPENDIX

in closing

Closing Comments

is to roll the credits and thank a whole bunch of people who made this all happen. Before I do that though, I'd like to take the prerogative of the chair here and just make a few additional comments from things that I'd written down in the last couple of days as well as some prethinking. One is the incredible speed with which we are moving ahead in space exploration. Now that sounds perhaps silly on the surface of it, but think for a moment. In the first 50 years of aviation, a million aircraft were built, most of them used multiple times. In the first 50 years of space exploration, there have been exactly 4500 launches total worldwide.

The difference, the gap, between where we are in commercial aviation today and where we are in space exploration is huge. The fact that Burt Rutan and his group can be so successful today is built on investments that were made, in some cases decades ago, by the government. Now where does this lead us? This leads us to establishing a viable space exploration industry eventually, such that there will be a trailing edge of people who can make a business case and make money out of not only communication satellites, but types of space travel.

Scott Hubbard
Director, NASA Ames Research Center

Scott Hubbard serves as director of NASA's Ames Research Center, in the heart of California's Silicon Valley. Hubbard's tenure at Ames began in 1987 and has included a variety of management roles. From 1997 to 1999, he served as the deputy director of the Space Directorate at Ames Research Center. Prior to his current appointment, Hubbard was deputy director for research at Ames. In March 2000, Hubbard was called to NASA Headquarters where he successfully redefined all robotic Mars missions in response to the Mars failures in 1999.

The two analogies that are already there are the railroads, and, as I said, commercial aviation. The railroads that got the right of way, Union Pacific, Southern Pacific, came together and drove that nail out there in Utah [joining the two railroads]. In aviation, the government invested in mail routes. And, eventually, this form of investment and technology and subsidies led to multibillion dollar industries. I think we are just on the verge of being able to see something like this come out in space exploration, beyond something like the commercial satellite industry. And I think it's going to be an absolutely fascinating journey over the next 10 years or so—maybe it's 5 years, we'll see—as this plays out.

The second major point is to underscore the false dichotomy of human versus robots. The only thing that will happen is the ratio will change over time. And, at some point, a human being, I'm looking at Chris McKay here, will be the tool of choice for exploring the Moon, and, particularly, exploring Mars. If we could put him in a little box, I'm sure he would go with the MSL (Mars Science Laboratory) in '09.

So where do we go with this dialogue? I agree completely with what John said, with what many of you said. This can't be a one-of-a-kind. I think the public will come along if we tell our story well, but we need professional help. Some people say we're beyond help; we need treatment.

But if we can talk about the risk of not exploring, the risk of losing our imagination, and maybe, ultimately, a second home for humanity, I think that we have some compelling things in addition to the kind of spinoffs that may come from what Nathalie Cabrol found by exploring these lakes, that your blood oxygenation goes up, your heartbeat goes down. What does that mean? What does that mean for the biomedical community?

There are a lot of things in there, but telling the big future story, I think, is something we haven't done and we need to do. And we saw some storytellers here in the last few days who just grabbed us. In giving a lot of talks, there's the pin-drop moment, and we hit the pin-drop moment in those places were everybody was just absolutely transfixed by the story.

So, where can we go with this dialogue? One thing is that taking risks can prepare you for the future—often in ways you didn't even think of. I'm going to give you one or two examples from my own experience, which has been largely taking programmatic and technical risks.

In 1975 very little was known about repairing neutron damage in gamma ray detectors. So I conducted, at Lawrence Berkeley Laboratory, a bunch of experiments with a 72-curie plutonium-beryllium source. Now, if you consider that your smoke detector is picocuries, you get some idea of how hot this was.

So, we went through the safety procedures. I signed up to the risks. Twenty-five years later, at the age of 50 or so, I had dual cataracts in both eyes, which was a possible outcome of doing that. But today, Bill Boynton with that same detector orbiting Mars, figuring out where all the water-ice is, is able to repair his detector, because of what we learned doing those experiments, almost 30 years ago, about how you heat the detector and get rid of the neutron damage.

You never know what kind of a risk, and what kind of information, is going to prepare you for the future.

In a similar fashion, in April of 1990, I went in front of the Headquarters folks and proposed the ridiculous mission of using a Delta-2 and a single probe and a cruise stage going to Mars and landing, of all things, using an airbag. The risk there was ridicule and being laughed out of the room, which almost happened. Fortunately, Jim Martin, the legendary leader of the Viking mission, thought there was something to it. And 14 years later, we have now used that technology three times successfully.

So, what I'm building up to is the analogy of setting up the Astrobiology Institute in 1998. We took a risk—and Keith Cowing was part of this—in bringing together an interdisciplinary group of physicists, biologists, mathematicians, astronomers, who never talk to each other. Or if they do, the intersection is only at one point. And saying, let's all think from our disparate points of view about

IN AVIATION, THE GOVERNMENT INVESTED IN MAIL ROUTES. AND, EVENTUALLY, THIS FORM OF INVESTMENT AND TECHNOLOGY AND SUBSIDIES LED TO MULTIBILLION DOLLAR INDUSTRIES. I THINK WE ARE JUST ON THE VERGE OF BEING ABLE TO SEE SOMETHING LIKE THIS COME OUT IN SPACE EXPLORATION, BEYOND SOMETHING LIKE THE COMMERCIAL SATELLITE INDUSTRY.

a much broader series of questions, like where do we come from, are we alone in the universe, where are we going?

Out of that came a field today, and you've heard references to it, of more than 1,000 scientists worldwide who are engaged in this in everyday research and view this interdisciplinary work, the interaction—the action is at the intersections—as being where we're headed for research in the future. So I would say today, the group that has participated the last two and a half days at this has been at a seminal, similar event of bringing together communities that have perhaps not communicated as much as they should—robotic, human, risk-evaluators, decision-makers. And, so, what we need to do to keep this moving is have the dialogue; perhaps we have a road map, we certainly need a distillation of lessons learned from this, and I would be willing to bet that we're going to ultimately have, if John takes his action items here, thousands of people, maybe tens of thousands of people, who are engaged in doing the kind of work that we got started here over the last two days.

So with that, let me roll the credits and, first of all, thank the idea men— John Grunsfeld, Keith Cowing, the people that had some of the initial concepts for this. Let me thank the Naval Postgraduate School, Admiral Dunne, and, particularly, Chris Walla for hosting us in this incredible venue. It's just been delightful being here. At NASA Headquarters, Bob Jacobs was the lead for pulling this together. Trish Pengra, Al Feinberg, and the inestimable Tony Stewart of NASA TV, thanks to you all.

The group from Ames, from my own Center, Rho Christensen, Danny Thompson, event coordinators. Victoria Steiner and Ed Schilling, public affairs. The video crew—I won't go through all the names. There are many, many people staffing the cameras here, but I do want to mention Jim Taylor and the planners collaborative, Mark Shaddock and Spotlight Productions, Donovan Gates, Donovan Gates Production, and Michael Ditertay and his staff on this 30-person television crew. And out of this will come, I'm sure, an outstanding DVD.

Then there are a couple of other people from Ames that I want to mention— Mike Mewhinney and Kathleen Burton of public affairs, who were part of the advance group getting all this together. Then, finally, a contributor, I'm looking at him right now—one of the real concept, idea, content contributors to this, who through some personal adversity, has managed to stay focused on making this entire thing very successful, Mel Averner. Mel, thank you. Then finally, our moderators—Miles O'Brien of CNN, Chris McKay, Dave Halpern, and again, John Grunsfeld, NASA Headquarters.

So, finally, to wrap it up completely, we want to thank all of you who have spent the last two and a half days with us, and, of course, the honorable Sean O'Keefe, the NASA Administrator.

NASA Image/Renee Bouchard

Remarks

after going through the day. I had given some time to thoughts about risk and the way we use it and the way we misuse it and all of those things. Well, as the morning went on, the first speaker ticked off the first three or four of my items, and the next speaker came along and fixed them all up. And, so, most of the things that I thought I would like to comment on were gone. Then, just to put the crowning blow on it all, we go in this afternoon and I listen to the most amazing set of people that I could ever imagine. And I'm sitting here listening to this and saying, "Every one of these people, individually, has done more than me and all my friends." Now, how in the world do you get up and talk after that?

Well, I decided that the first thing I had to do was to talk about something different. So, what I would like to do tonight is perhaps a little deviation, but I hope my thoughts are in the context of what you are discussing.

We have a nomenclature issue when we talk about exploration and the word explore: to some people, that means visit planets. To some people, it means do great science. To some

Thomas "T.K." Mattingly II
Former NASA Astronaut

T. K. Mattingly is one of the 19 astronauts selected by NASA in April 1966. He served as a member of the astronaut support crews for the Apollo 8 and 11 missions and was the astronaut representative in development and testing of the Apollo spacesuit and backpack (EMU). He was designated command module pilot for the Apollo 13 flight but was removed from flight status 72 hours prior to the scheduled launch due to exposure to the German measles. He has logged 7,200 hours of flight time—5,000 hours in jet aircraft. A veteran of three space flights, Mattingly has logged 504 hours in space, including 1 hour and 13 minutes of extravehicular activity (EVA) during his Apollo 16 flight. He was the command module pilot on Apollo 16 (16–27 April 1972), was the spacecraft commander on STS-4 (26 June to 4 July 1982) and STS 51-C (24–27 January 1985). After retiring from NASA in 1989, Mattingly continued his work in space science in the private sector, focusing on developing low-cost and reusable launch systems for commercial use.

of us—I like to call ourselves explorers—but I tell you what, the ride is one hell of a good show. So I think we have different perspectives on what exploring is.

But once you get away from this community of ours, you find that the word takes on a different connotation. We use the expression "to explore business opportunities" and the expression "explore new kinds of things." And, in my mind, this "explore" means to do or to learn something new. It doesn't matter whether it's science or how to manage an organization or how to go places. It's when you do something new. And, in my mind, that can take on something of a different connotation. And so, if you look at it that way, then there are a lot of people in the world that take risks. In our business, we talk about risk and the first thing we think of is some poor kids' young bodies laying there in the ashes.

Well, there are a lot of other risks, and they're very, very real and they're very important. And for those of you that have tried to start a business or have tried to run one with your own money, you understand what the word risk means. And it is just as overpowering as anything else.

When I had an opportunity to launch the Atlas rockets—which, by the way, I consider to be one of the highlights of my career opportunities—I can tell you, it's infinitely easier to sit on top of one of the things that NASA launches, because you are absolutely in the best hands you could ever be and you will never find a lower level of risk. When you go launch it and it's your decision—it's not a committee—you've got investors that you've just assured it's going to fly. But it's the same old rocket hardware and it's just as interesting. And that really gets your attention.

I know that the docs like to record the heart rate—they want to know what Jim's heart rate is at launch and at entry and when he steps around. I tell you what—any of those statistics they collected on us won't compare with making the decision to launch something that's got your money riding on it. That's a different ball game. And it is just as interesting as people.

So, my point is not to belittle people. My point is to say risk is a different thing to a lot of different people for a lot of different reasons. And, so, when we say we're taking an acceptable risk or whatever we're going to do, you have to put yourself in the place of "risk to whom and for what?" One of the speakers this morning reiterated that we all think of risk of life. Okay, that's pretty easy. There is a property risk, but, actually, I think we can take almost all physical property and lump it together under financial arrangement of some sort, except in those rare cases when we're going to use or deplete a natural resource that doesn't get refurbished. I remember one time in the Shuttle program, we just woke up one day and discovered that our demands—if we met the flight schedule—would have depleted the Earth's supply of helium the first year. So we kind of had to do some more engineering. So there is an example of another kind of property that you put at risk. But you also put at risk opportunities, and that's opportunities for you to do something else with your time. The investor could invest in something that's going to come out better—there's a million things that could happen. So the connotation of risk is something that you have to stop and think about.

It seems to me that in a democratic world one of the principles we have is that there are human rights that belong to everybody and we go to great lengths to take care of those. And as school kids we were taught that our rights would end when yours start. Okay? That was an easy principle.

That same thing applies to third parties. When we do our trick and we launch things over people and around them, or when you run nuclear power plants, or when you do all kinds of things, there are innocent parties who did not get to vote on taking a risk. And one has to think very seriously about who it is that has the authority to put in jeopardy somebody who didn't even participate in the decision. One of the nice things about the discussion this afternoon was everybody that I listened to was in activities that did not put third parties at risk. They were responsible for other people, they were responsible for a lot of things, but the innocent bystander was generally immune to their activities. And so that [responsibility to innocent parties] puts an obligation on all spaceflight from the beginning.

When we go out of the atmosphere on those missions and come back in, we're going through something that's very traumatic and irreversible. Spaceflight is complex by its nature. It's large in scope and it has a whole range of critical, irreversible decisions in a harsh and unforgiving environment. Other than that, it's a wonderful place. [Laughter]

That first step has got to be right, and with that comes an obligation to all those kids out there in the world that aren't part of our club and aren't having fun doing things that we enjoy. It's easy for us to decide, "Hey, this is good stuff," whether it's good science or just a ball to go do, that's one thing. That's different than saying, "I'm going to fly over your cow pasture and maybe drop something on your house." People tend to get irritated at that.

So, what I wanted to do is step back for just a second and talk about some of the perceptions so that it can help frame the question. Now, I'm not a visionary. I don't know what the world should do—I don't have any idea about whether we should explore Timbuktu or Saturn or whatever. But in my opportunities in life, I've had a chance to do a lot of really neat things where you could have a vision about how to get it done. And so I guess I'm one of those people you call an implementer instead of a visionary. That's what I enjoy doing, and I think that's the kind of things that have just worked out in my favor.

So while not a visionary, I have watched some. What I'd like to do is share with you some thoughts about groups that I have watched and the characteristics of them. Because I'm going to make two assumptions—and these are not debatable, because they're assumptions. I'm going to assume that you either go forward or you die. Civilizations do that. So if you aren't making progress, you're in deep trouble. Maybe it'll take time to play out, but that's the end. And I can't prove that, but, boy, do I believe it.

Somebody gave me the analogy that it's like riding a bicycle. If you try to sit still and not move, it's a very difficult job. And if you can get up a little speed, you can do a lot of things. That's one assumption.

The other I'm going to assume is that there is no way we will not explore the universe. I have no idea what the timeframe is, but one of my investor friends gave me a piece of counsel one day when we were having trouble and couldn't figure out how we were going to make the next step. And he said, "Just don't get in the way of success."

Going back to Jim Lovell's [Apollo 13] mission, there's one lesson that I gathered from our ground risk management and getting a chance to watch the real pros go do that. When we started, within an hour of Jim telling the world he's got a problem, we didn't have electricity, we didn't have oxygen, we're on a trajectory that's not coming home, and we don't have any ideas. And those cats on the ground solved these problems one at a time. The only rule was, you've got a problem to solve, you've got one to solve, and you've got one to solve, and we do have a cutoff date when we need to have all this finished—it was later than Jim wanted it, but it beat the deadline.

But the principle was, don't get in the way of success. Assume that your buddy is going to do his job and you don't want to be the one that's holding up the show. With that, we went through a series of really challenging resolutions to problems. Where folks really didn't know, but they said, "Boy, if they can figure out how to get the water to last, we'll figure out how to get the electricity over there." And it all came together, as you know.

So I'm going to assume that we're going to go do these things and that we're mature enough we recognize that, I think, every success is preceded by a failure. At least in my experience, it's not real clear you can have a success without preceding it with something that's humbling or threatening. Certainly my career has gone through that sort of cycle.

The things we learn, we learn most easily from things that don't work. You've got to be objective, you've got to be honest with yourself, but the things that fail are the things that teach us. I have known a few people who could learn from success, but you know, when you're feeling good, it's really hard to be self-critical. And so you miss a lot of lessons that you could have had. So don't ever be afraid of failure.

So, if that's the case, if my premise is right—we're going to make progress and we're going to go explore—then our job is don't get in the way of success. We don't know from the government side what the funding profiles will be, what the timing is, but we need to be prepared to do whatever opportunity presents. So how do you do that? I don't know. And I certainly wouldn't tell you anything other than sea stories about places I've been. But we're not in those places. We're going forward. And that's a new game and a new set of challenges and new places to go. That means rethink.

So, in that vein, let me just summarize my observations from spending 20 years in government programs and then a few years working as a contractor on government programs and then the last ten years working on commercial ventures. I've been in large corporations—government certainly is a large organization, DOD's a large organization—and I've been in some small startups and entrepreneurial ventures. We've made mistakes and we've had some successes. So I've tried to catalog for you the signatures that have shown up in every success. And some were hinted at today and I just wanted to reiterate them.

Number one, you have to have a clear, quantifiable, simple-to-understand objective. Step one. If you don't fill that square in, don't worry about the rest of them, because they don't matter.

Once you've got that, you have some more challenges. And it takes creating an environment where getting it right is more important than who's right. You have to have a group—and big things can't be done by small groups and by individuals, only by large organizations. The trick in leadership is to create the environment where getting it right is all that counts, because the job's too hard to do anything else.

NUMBER ONE, YOU HAVE TO HAVE A CLEAR, QUANTIFIABLE, SIMPLE-TO-UNDERSTAND OBJECTIVE. STEP ONE. IF YOU DON'T FILL THAT SQUARE IN, DON'T WORRY ABOUT THE REST OF THEM, BECAUSE THEY DON'T MATTER.

So if you've got that, then you have to have competent practitioners. Without that, you won't go anywhere. Now, back in the Apollo days, that was one thing no one had to worry about. Because if you just said, "Job opening—work on Apollo," you know, the line went all the way around the county, because it was something every one of the young kids wanted to do.

Today we have to compete for opportunities and people, especially. They will come to an electric environment. The kinds of things that you folks do will draw people. They are there—and they're the people who want to be there, people who want to be personally accountable.

So in this group, this constellation of things that I have observed as uniform qualities, you have got to have a good objective, you have got to have personal accountability—eyeball to eyeball, participant to participant. That's not an org chart with lines on it, that's real-world accountability based on human relations that we have with each other. You have to be competent in your job.

I would caution that one place we've gotten trapped is the resume trap or the logo trap. I'm the world's worst in reading a resume and knowing what somebody can do. I feel pretty good after working with them for a couple of days, and then I know what kind of people I'm around. But I have a real hard time with a resume; they can look really good or really bad.

The logo trap is the other side of that. How many times have we worked in an industry that's maturing, where the logo of the company is on the wall and it has a record of miraculous accomplishments—year after year they've done spectacular things. All of us, including the employees, believe that we are part of that logo. And it happens at NASA, it happens at any large organization with a history. We identify with that logo, that's a symbol of things that have happened.

Maybe, after a period of time, the people that do those things aren't there anymore. And unless somebody has been very, very careful to be prescient enough to create an honest-to-goodness succession plan, you'll find people who know the language, who look good, but do not have that personal, gut feeling for what it's about that's necessary to do these things that push the envelope. When you find that situation, the places that succeed recognize it, and then they take steps to fix that.

There's nothing magic about this except to face up to the fact that you know what you know and you know what you don't. And with that, those signatures have shown up at every one of these little organizations that I've had a chance to be exposed to.

So, while I can't tell the answers to the next job and the next challenge, because each one's unique, I would commend to you that these observations, that I think I picked up primarily from working at NASA; they have been uniform signatures. We even applied the much-maligned aerospace management process to turning around a very nonglamorous company, where we did a really excellent job of turnaround, coming out of bankruptcy to create some almost embarrassingly good results—done with people in a nonglamorous field and a group that two years ago was absolutely demoralized and hopeless.

It all came from just getting them all on the same page with the right orders. So these are techniques that are not just peculiar to the high-tech business, they work everywhere in life. So that's my observation.

I do have one question I'd like to ask of you. When I was a kid, I lived in Miami, and I used to go down to the beach, like all high school kids, and look up at the sky and see the Moon, and you kind of wondered, "Gee," you know, beer talk, "Hmm, wonder what the Earth would look like [from] up there?" Well, that was too preposterous for even high school kids to talk about. Strange things happen.

I had a chance to go and serve what I thought would be a couple-year tour with NASA, and they were doing this program called Apollo and space-centered life. I knew that when I got there, I wasn't going to the Moon. But, you know, I might be getting in at the right time to go to Mars. [Laughter] Well, that schedule has been modified a couple of times and I said, "Well, okay. I did get to go to the Moon, I hope that doesn't blow my trip to Mars." [Laughter]

Then I woke up and said, "Maybe I could be the program manager to send somebody to Mars." So tonight, I would plead with all of you in the exploration world. Before I turn the lights out, I want to see pictures of people bouncing on Mars. And that's your job. Thank you.

Remarks

...awful lot of eye-opening stories about how we are exploring the planet today. I'm awed to be in the presence of so many notable people here in the Monterey Bay Aquarium. Actually, this aquarium has figured inspirationally in motivational movies such as *Star Trek* and other grand works of science fiction. I should tell you I have been motivated by *Star Trek*, I think we heard that this afternoon from Chris McKay.

Preparing for this talk, I continually asked myself why I, of all people, have been asked to speak to you this evening. And I kind of went through the thoughts. Maybe because I most recently returned from space—that seems an obvious one. Or because I've been fortunate enough to survive six flights to space. Or worse, because somebody sees me as prone to avoiding near disasters throughout my life. I know someone in my management chain believes that.

I do not feel I'm a particular specialist in risk-taking or taking risks personally. Rather, I see myself as rather conservative about mitigating risks that I see ahead of myself and my family.

Michael Foale
NASA Astronaut

Michael Foale was selected as an astronaut candidate by NASA in June 1987. He served as a mission specialist on STS-45, STS-56, STS-63, and STS-103. He was flight engineer 2 on *Mir* 23 and *Mir* 24 (ascent on STS-84 and return on STS-86). On his last flight, 18 October 2003 to 29 April 2004, Foale served as International Space Station (ISS) Expedition-8 Commander. The Expedition-8 crew launched from Baikonur Cosmodrome, Kazakhstan aboard Soyuz TMA-3 and docked with the ISS on 20 October 2003. His six-month tour of duty aboard the International Space Station included a 3 hour, 55 minute extravehicular activity (EVA). Mission duration was 194 days, 18 hours, and 35 minutes and, at its conclusion, Foale became the U.S. record holder for most cumulative time in space, having logged 374 days, 11 hours, and 19 minutes.

There are many guests amongst us who do not work at NASA, but have very relevant experience in exploration. Please believe that I see risk perception and its mitigation as a rather subjective issue—I think we've heard that a number of times today.

I, and NASA, do not know all the answers. In fact, I feel we may have strayed off course concerning our approach to risk in some areas. We, NASA, need to hear more than anything else not Mike Foale's point of view on risk, but those of people outside of NASA looking in. I feel my job today is to sort of set the scene and issue provocative opinions to you—I mean, I'm opinionated—and you are obliged to dispute them in the coming days.

That said, I'm going to give you my personal view of America's space exploration and the risk that comes with it. But first, I'd like to set the scene for space exploration in the future, inspired by the President's vision for exploration, by showing the first part of a video made within the astronaut office by astronauts and narrated by astronauts—one of whom is myself.

[Narration from video is indented.]

Female speaker: We are, by nature, explorers. Look at the centuries of histories where people were committed to finding new worlds and establishing them. And now I think it's time for us to go beyond low Earth orbit and do the very same thing.

Female Speaker: Human beings are insatiably curious. We want to know what's out there in the stars. It's part of who we are; it's part of what we are.

Male Speaker: Being outside on a spacewalk is the coolest thing you can imagine—beyond belief. You're doing this important thing, you're building a spaceship and the world is rolling by. It's absolutely breathtaking.

Male Speaker: The Space Station is teaching us how to explore. Before we can go to the Moon or to Mars, we have to learn a lot about the human body. What happens when you put yourself inside a spaceship for weeks and even months? What food are we going to eat? Are we going to bring it all in cans or are we going to grow some food on board? What sort of spaceships do we have to build?

Michael Foale: When we look back 50 years to this time, we won't remember the experiments that were performed, we won't remember the assembly that was done. What we will know was that countries came together to do the first joint international project, and we will know that that was the seed that started us off to the Moon and Mars.

Male Speaker: I think you have to learn to live and work on the Moon first, so you can make mistakes when you're only two and a half days away from a can of beans.

Male Speaker: Human beings can do things that robots will never be able to do. They can anticipate, and they can handle, unexpected problems.

Male Speaker: On the Moon, we ran into about 97 problems that nobody thought we'd run into, and we fixed every one.

Male Speaker: We are going to continue to explore. We can confront the majority of the problems by going to the Moon. And then, building on that will give us the confidence and the technical ability to be able to step further into the solar system and turn our sights towards Mars.

Male Speaker: We go to places where human beings typically can't live because these environments offer discoveries that defy our imagination. We're going to say, "Wow!"

Male Speaker: We want to know where we should land; we want to know where the water is. The robots blaze the trail—provide us with a path to get there. They're finding out whether we could stand on the surface of Mars. Those robots have raised their electronic eyes and given us those first glimpses of the horizon of Mars. To be able to stand on the surface of Mars and feel the wind blowing of Mars's thin atmosphere is going to be a tremendous achievement.

Female Speaker: Can we use some of these resources? Can we prosper here? Can humans live here?

Male Speaker: So far, we have only sent people as far as the Moon, and sent our robots just as far as the edge of our solar system. We are just starting to understand our place in the universe, the perspective that the universe gives us, and the tremendous, infinite variety that the rest of the universe holds. That's where we are headed, and that's where we'll go after Mars.

[Video segment ends]

After watching that video, or others just like it, I find myself kind of naturally responding with enthusiasm and excitement. I kind of go, "Wow!" It makes me feel that we humans can do anything if we agree on a common purpose and simply put our minds to it.

However, evocative and inspirational as my astronaut colleagues can be, we are leaving out of the message something terribly important—risk. Why is that? It's because we feel instinctively, maybe—especially in this year—it will spoil the mood of our message. That it will conjure up very painful and recent memories of lost friends and failed missions.

My theme to you this evening is that we must always talk about risk when we enthuse about exploration. The two are inevitably connected. And I think that message is coming home today.

Risk—what is it? It's obvious when disaster strikes, such as when Shackleton's ship, *Endurance*, was forestalled in his second attempt to reach the South Pole, crushed by the ice while trapped far from his goal. We consider an activity to have risk if a foreseeable outcome has undesirable or dangerous consequences. Everybody knows what risk is, but it's according to their own subjective standards. Risk today, in Western society, might be perceived to be—as T. K. Mattingly referred to—a financial activity or the stock market, allowing your children to take the bus to school, not evacuating in advance of a hurricane warning, or not wearing a seatbelt. And these examples are seen as risks because the consequences can significantly change our lives through financial ruin or loss of life.

So this evening when I speak of risk, I mean the risk of people being killed. Historically, or even today in underdeveloped countries, loss of life was an unfortunate, but commonplace, occurrence within families and all other types of social units. Every child experienced soon in their childhood somebody dying or they saw a dead person. This might have included the ravages of marauding neighbors, war, starvation, and disease.

Before Christopher Columbus, if a proposal of exploration was made—be it to scout the far hills and tribes at a distance, or to utilize substantial resources of the community to send ships on marauding or exploring adventures—the risk entailed would appear to carry consequences not worse, and possibly better than, the risk of inaction.

Inaction might simply mean waiting for unknown peoples to find and attack the community or running out of food or tradable goods. So the imperative to explore then and to take risks then was strong, because the risk was understood widely to be a means to survival and the reduction of future risk.

When a ship that had carried away a large fraction of the able-bodied community did not return or became known to be lost, the news would be just as painful then as it is today, but I think the shock should have been less.

How do exploration and risk play a part on Earth now? I see exploration taking place under the sea, such as underwater archaeology, or on land, such as the search for Mars meteorites in deserts or Antarctica, or in mountaineering— and in space, as we develop human and robotic space missions beyond the realm of Earth. I do not see these combined exploration activities consuming anything but a small fraction of the world's economic and human production.

I do not know how today's activity should be compared to that [of] more than a hundred years ago, but my feeling is that outlays for exploration today represent a smaller fraction of our output than in the past. So, in risk terms, nowadays activities are just as dangerous for participants as any exploration undertaken in history—dying is dying. There has been no change in the fact that people can be injured today and lose their lives while exploring.

What has changed is the public expectation for success, and the public shock when risk and danger show themselves as injury and loss of life. We're not often exposed to death and severe illnesses or injury in our personal lives, unless we're in a group that we could label as thrill-seekers—and we've been avoiding that term here today—or work in medical or emergency services, or in a war zone.

I'm going to show you slides of a series of missions that I did not take part in. I was too young. I was just an enthusiastic, dreamy watcher of these events that took place in the '60s. I'm going to show you astronauts walking out to their vehicle and then the vehicle launching. And I want to tell you to think about how you, the manager sending that astronaut out to the launch pad, might feel—or the family. And then I think about how you, as the astronaut or the risktaker walking out to that launch, might feel about your risk.

[slide] This is Alan Shepherd getting into his Mercury capsule in 1961, May 5th. After the Soviet Union had orbited Yuri Gagarin, April 12th of that year, President Kennedy stated in a press conference, "No one is more tired that I am in seeing the U.S. second to Russia in the space field." And he went on to say,

WHAT HAS CHANGED IS THE PUBLIC EXPECTATION FOR SUCCESS, AND THE PUBLIC

SHOCK WHEN RISK AND DANGER SHOW THEMSELVES AS INJURY AND LOSS OF LIFE.

"We are, I hope, going to be able to carry out our efforts with due regard to the problem of the life of the men involved this year."

So he did not say it directly, but he was referring to the high risk of putting a human into space. James Webb, the then NASA Administrator, issued a statement no more optimistic. "NASA has not attempted to encourage press coverage of the first Mercury Redstone manned flight." I think that's incredible in today's environment. "We must keep the perspective that each flight is but one of many milestones we must pass. Some will completely succeed in every respect. Some partially, and some will fail. From all of them will come mastery of the vast new space environment on which so much of our future depends."

[slide] This is Alan Shepherd's lift off on a Redstone rocket, flying for no more than 15 minutes until splashdown. The flight was a success. Afterwards, the risk perceived by the public may have been assuaged a touch. But my point to you is, because this was a first flight of a new nature carrying a human, it had great risk. So, like a test pilot, I believe any first flight with a human being carries increased risk, especially in recently designed, new space vehicles.

I'm going to show you a series of slides of space missions, as I mentioned, that I believe carried a particularly high and increased risk. Initially, these missions are ones I did not take part in, and so your opinion is as strong as mine. I think you should hold your opinion and see if it corresponds with that which I'm going to express to you.

In some cases, this risk may have been well understood by the public, such as this first flight of Al Shepherd. Other slides I will show, the public was much less aware of how great the risk was and found themselves surprised. [slide] Here's John Glenn, 20 February 1962, walking out to the first human flight of the Mercury-Atlas vehicle. John Glenn walked out to a much more risky launch than the one before him by Gus Grissom, which had also been on a Redstone rocket. Why? In my opinion, it's pretty clear. Because the vehicle had been changed. The mission was very different. Launched to orbit with 3 times the speed of the Redstone, 10 times the energy to gain getting into orbit, and 10 times the energy to dissipate in excess heat reentering from orbit.

This is the basic fact of the physics of spaceflight into orbit and away from the Earth. The energies needed to be acquired or dissipated are huge, roughly 300 times the kinetic energy of airliners, 290 that of supersonic jets, 25 times that of *SpaceShipOne* this week, on which I, personally, pin much hope, and I think the rest of you do, also.

Was this huge difference compared to Alan Shepherd's flight understood by the public? Kennedy did say only later that year, in September, "We choose to go to the Moon in this decade and do the other things not because they are easy, but because they are hard."

[slide] When Gus Grissom and John Young walked out, in March 1965, to Gemini 3, the risks were again increased, in my opinion. It was a new human launch vehicle, a first flight for humans, and it was a new, larger spacecraft, the Gemini capsule. On the previous rockets, there was an escape tower. The crew escape system was reduced in this case—ejection seats—diminishing its capability compared to Mercury. It was a big, risky step for our nation's space program, but probably not perceived [as so] by the public.

[slide] This is Ed White on the first U.S. spacewalk—definitely a new risk in our space program, adding to others as a first-time test.

[slide] Here's Neil Armstrong and Dave Scott docking with the Agena upper stage, only to experience high rotation rates when they docked. They undocked and experienced even worse rotation rates, tumbling. They saved themselves by switching to a different attitude control jet system and made an emergency splashdown thousands of miles from the planned recovery area.

So the risk of human space exploration then, in this program up to that point, had been successful. Shows itself as a real hazard, but in NASA parlance, we call that a close call. It's where we go, "Whew! That was dangerous," breathe a sigh of relief, but nobody lost their life.

[slide] The death of the Apollo 1 crew—Gus Grissom, Ed White, and Roger Chaffee—in January 1967, in a fire inside the command module while on the launch pad, pulled NASA and the Nation up short. But the tragedy brought the best out in NASA and the Nation at that time, with new public resolve and tough lessons learned.

[slide] Two years later, an incredibly bold and risky decision was made by George [Mueller] and others to send Apollo 8 to the Moon after only one manned Apollo flight. Jim Lovell talked about that this morning. I think it is an incredible flight, especially risky because they did not take a lunar module with

them, which, because of its independent systems as a spacecraft in its own right, mitigated for future Apollo missions the risk of command-module failure.

[slide] Apollo 11 was well-perceived by the public to be risky. I think failure would have been tragic, in their minds, and awful, but not a shock. There was the unknown risk of landing on the lunar surface, plus the high risk of the Apollo system as a whole, but, so far, successfully flown. I remember as a young boy of about 12 or 13, the success made me sigh with relief, as if the risk had somehow gone away at that point.

The reward for the United States, for the Nation, when we are willing to take risks and to explore, is really so obvious in lunar rendezvous; the liftoff from the lunar surface with just one engine—only one engine to get you into orbit—carried a whole other set of risks with it.

And then we come to Jim Lovell's flight with Apollo 13. Its emergency was more of a type—in my mind, Jim—that NASA actually expects and tries to plan for. Risk again showed itself as real. I've wondered how I might have felt leaving the Earth when the accident happened.

As he pointed out, it was a fortuitous place—200,000 miles from the Moon—from his point of view. But, in my case, I think of not being able to turn around as the power systems of their command module failed. I think of what the cold, dead spacecraft may have seemed like when I was on *Mir*, when we lost energy, lost power, without a single sound and no power and the cold of space sucking the heat out of the spacecraft and yourself and your crewmates. It's a very, very hard task dealing with a dying spacecraft because it gets so cold and so wet. For Apollo 13, the risk was seen to be a close call. I don't mean to diminish that, Jim Lovell, but it was a close call because we pulled it off—you pulled it off—no one died, thanks to thousands of people on Earth and your crew.

[slide] STS-1, with John Young and Bob Crippen. This was the first powered flight of a Space Shuttle. I feel this was the boldest, riskiest flight in NASA's history. But if you mention that to John, he just seems to mutter some understatement characteristic of only John Young. The launch involved three characteristically different components to work perfectly and all together for the first time in a manned test. These were the external tank, the solid rocket boosters, and the orbiter. And within these, the main components—engines, hydraulic power units, fuel cells—all had to work reliably, but at least these had been tested in an integrated fashion before powered flight. This was not true of all three components together. No unmanned flight of the STS had been conducted. And the buildup to STS-1 was slow and difficult for NASA, so the public heard about its risk in the press as much because it had been so long since the last manned launch of Apollo to Skylab. For all that risk, the crew escape system—ejection seats—was especially limited compared to that of Apollo, adding even greater risk to the crew for this first flight.

But STS-1 was a success, as were subsequent flights up to the 25th, *Challenger*. The ejection seats were removed. Our public and NASA seemed to expect space exploration to be like that of airline operations. And to be fair to the public, this is an understandable misconception.

261

Only recently—just two weeks ago when I was climbing Mt. Baker—we were discussing the loss of *Columbia* with people who do not work in the space program. And the genuine question goes, "After all, the Shuttle lands like an airliner, right? So it must be as risk-free as an airliner. You spent all that money on it." I've heard this from generally well-informed people in different professions. So the public is especially shocked when the Shuttle is destroyed.

Okay, so why do astronaut applications to NASA actually increase after we've had a disaster, including me in 1981, watching STS-1 from Cambridge, England, driven to become an astronaut. Would-be astronauts do risky things to acquire the skills of explorers—I think Bill Stone overdid it this morning—such as fly gliders or scuba dive on expeditions in Greece; this is something I thought was really captivating and interesting. Or excavating human remains in the low visibility and cold conditions on the *Mary Rose* in the English Channel. There was risk in these exploration activities for me and the two people who preceded me. Two people had died in the course of many dives on the *Mary Rose* [before I joined the] project. But the excitement of discovering new things was compelling and it pushed me to do more.

[slide] Becoming an astronaut in Group 15 in 1987, after *Challenger*. Yeah, you'll recognize some characters here, it's an in-crowd, but it was a result of my desire not to take risk, but to experience space exploration. My desire outweighed the risk I perceived, a risk greater than I probably realized at the time.

[slide] This is astronaut spaceflight readiness training, and it carries risk. We may have to eject out of a T-38 or be picked up by helicopter in search-and-rescue exercises. Or—this is not hazardous—overeat during a survival exercise. But these training activities to prepare astronauts are undertaken to reduce our future risks during space missions.

So our training carries risk also, and this is to be balanced carefully with the higher risks that we are trying to mitigate in the conduct of our space missions. Our remote outdoor expedition training is a key to preparing crew members to make use of local resources, solve technical and mechanical failures in difficult conditions.

[slide] Here John Young and Charlie Duke are being trained in geology to increase the science return of Apollo 16, which was highly successful. I believe we need to place future exploration astronauts into geology field work, in a long-duration expedition context, as part of scientific expeditions where scientists have a stake in these activities of the astronauts. So the astronauts feel the pressure that stake has on them, [as] for example, searching for and recognizing Martian meteorites in the deserts or in Antarctica.

Post-*Challenger*, my first flight was on STS-45 in 1992. And my family took the risk very seriously, as the families of all astronauts do, as did my first commander—Charlie Bolden. And he was already a three-time flyer, I think, at that point. And he strongly encouraged me—and I was a bit surprised by this—to write a will. It was honest advice for a risk-taker from a risk-taker.

NASA managers work to the very best of their ability to manage our risk when we fly, but they are limited to the tools at hand, the architecture of the

Space Shuttle system, and the inherent risk in all launch systems attempting orbital speeds.

In the late 1990s, NASA was directed to work with the Russian Space Agency to build the International Space Station (ISS), providing sustaining financial support to, at that time, a Russian space industry in severe difficulty. And it jump-started the redesign of the ISS and initiated a series of joint Shuttle-*Mir* missions throughout which a NASA astronaut would be left aboard the *Mir* to gain experience in the conduct of long-duration space flight.

[slide] So here a few of us and our Russian support staff are gathered in front of Yuri Gagarin's statue in Star City. As Charlie Precourt and our crew brought me towards *Mir* in 1997, I was anxious actually not about the risk, not for my safety, but my ability simply to interact well with my Russian hosts, my cosmonaut crew. The launch was behind me, and I reckoned the on-orbit phase should be less risky.

Lloyd's of London must have thought the same, because they charged me the same $1,500 for mission life insurance, just as they had for my shorter Shuttle missions. They would have been horrified as that mission unfolded, I think.

The risk of the U.S. working with Russia in the conduct of these expeditions was that the two sides did not, and could not reasonably, know everything about

LLOYD'S OF LONDON MUST HAVE THOUGHT THE SAME, BECAUSE THEY CHARGED ME THE SAME $1,500 FOR MISSION LIFE INSURANCE, JUST AS THEY HAD FOR MY SHORTER SHUTTLE MISSIONS. THEY WOULD HAVE BEEN HORRIFIED AS THAT MISSION UNFOLDED, I THINK.

each other's decisions and processes. I certainly did not know or understand that well at the time. A lesson learned during this program was that we are obliged to know as much as possible about each other's operations that carry risks.

Jerry Linenger, who I was replacing, happened to tell me in the handover a hairy story about a manual Progress docking attempt, which Vasili Tsibliev had been instructed to carry out earlier and which, in the end, failed, finishing in a close call, a fly by of the station. I listened attentively, but did not know how to calibrate it as a risk. At any rate, I considered the presence of an independent space vehicle—the Soyuz—to be sufficient to insure our lives in the event of bad events on the space station. And, as it would turn out, we came very close to testing my supposition.

I'm going to show you, very briefly, a clip of a collision of a Progress vehicle that took place while I was on board the space station *Mir* in 1997. Before the actual collision takes place in this video, I will show you the way this docking attempt should have taken place. There you will see a Progress vehicle coming in towards the space station, towards the docking axis. And it will dock in a nominal fashion, stopping at about 100 meters, and then the crew takes over, using manual controls.

In this successful attempt, carried out by Anatoly Soloviev and Pavel Vinogradov, that I witnessed actually later on in that year, they were using all of the full capabilities of the Progress docking system—the range and range rate, the radar system—that allow a normal automatic docking to take place. Vasili Tsibliev, my commander in *Mir* 23, had been asked to turn off that equipment—not use it. Why? Because the program in Russia wanted to cut the cost of buying a $2 million electronic box in the Ukraine. That was the rationale for this test. As it unfolded, and as I learned about it, I realized this was a gross miscalculation of what we were ready to do that day, and it was very improperly thought through [about] how to carry out this docking test.

[video] The sound you hear is in the Soyuz as I was flying around in there looking at the damage, actually. This is the docking module that we're talking about—the docking core. Here's Anatoly monitoring the TORU docking equipment. And he sees the *Mir* in his sights as he flies the Progress manually, looking through a camera from the Progress towards the *Mir*.

This now is the scene as Vasili saw it. We'd already gotten too high above the *Mir*. You can see the solar arrays of the *Mir* here, this is the long axis. I snuck this video, by the way, which is why it's such poor quality. They didn't know I took it. And the docking was along this axis, it was meant to be. You can see we're high above the *Mir*. Vasili is not really saying anything in this audio yet. I'm just watching over his shoulder. Sasha is nearby. We should be docking on this axis, but we're now moving this way.

Sasha is saying, "You should move out." Sasha is saying, "Break out! Break out!" He says to me, "Get to the spacecraft." This is my feet coming by the scene here. And then, that's the crash as the Progress hits. At this point, I'm floating into the space towards the Soyuz and the pressure's already falling, I can feel the pressure, in my ears falling.

This is the classic klaxon that you hear when you have a loss of pressure. Afterwards, when we did the survey, flying around in the Soyuz spacecraft, we looked at the damage and we saw that the solar array had been badly crashed.

After big events—after risk—you relax. And I wanted to show you what the handover's mood looked like as we finished up. After that pretty terrible day for Vasili Tsibliev and the rest of us, but particularly bad for the commander who suffered the stigma of this collision, every day we would look out of the window at this scene.

The damage to the Spektr module was serious, and it broke the foundation of that solar array that comes in here towards the Spektr module—so much [so] that I feel that the bearing was the location of the breach in the hull or [the] leaks. And Anatoly Soloviev and I did a space walk in Russian suits to survey the damage and try to find a hole, but we were not successful.

More serious and risky were the successive—and this takes me back to Jim Lovell's experience—times when we would lose complete attitude control of the space station and tumble slowly. When we had isolated that module—the Spektr module, Sasha and I—after the collision, we had cut off 30 percent of the *Mir's* power supply in so doing. And so now, the *Mir* was in a very critical energy state.

Actually, orienting the *Mir* using the Soyuz, which was the way we did this to overcome the loss of attitude control, always made me nervous that we would have inadvertently stabilized it in a spin, so stable that we would forever be stuck in it and direct the arrays away from the Sun and then, therefore, kill the station.

[slide] This is for John Grunsfeld. To put risk on *Mir* in perspective, I have to add that the risk of a Space Shuttle flight, for me, after the *Mir*, was just as real to me. It was while participating in a Hubble repair mission—with John Grunsfeld, by the way, over here—on STS-103 in 1999, commanded by Kurt Brown, that I felt the most anxious about what we're planning to do. And the task simply was performance anxiety for me. To change out the brain, the main computer, of the telescope—that made me more nervous that day, about my own performance and the risk of my actions, than anything I have ever experienced in all of six space missions. To leave Hubble worse off than we had found it, now that was a nightmare I did not ever want to contemplate.

Coming back to Russia again, NASA's experience on *Mir*, I believe, went a long way to reducing risk in working with the Russians on the International Space Station. We gained insight into their commissions and launch decision-making processes.

[slide] So here you see me. I want to show you, this is the management point of view, and it's a serious one of launch readiness. Ten days before launch on that Soyuz TMA-3 in October of last year, I am being presented as kind of an item—Exhibit A—to the Russian commission. Not only as a risk-taker, but as a form of risk mitigation. The argument was presented, in front of me and my crew, by Star City that our training was complete and sufficient and so, therefore, our performance did not represent a risk to the completion of Expedition 8. It was kind of a unique situation to be in for me.

As we approached the time of departure from Star City to the launch site in Baikonur, Kazakhstan, my family—Rhonda, Ian, and Jenna, and those are my crewmates, Aleksandr Kaleri and Pedro Duque—were toasted very seriously by the Russians and thoughtfully, acknowledging the unspoken risks in front of us as we embarked on Expedition 8. At this point, no one talks about risk.

[slide] I'm going to show you the walk out from the suit-up building in Kazakhstan out to our designated squares, and then the salute, and then on to the launch pad for a Soyuz launch. On the way out to a Shuttle launch, you become introspective, somewhat, as you notice all the other vehicles for a Shuttle launch are leaving. On the way out to a Russian launch, I'm always amazed that in Kazakhstan, when you get to the base of the rocket, you're surrounded by

hundreds of senior figures and VIPs, and they're all clamoring to be there, right next to a steaming, hissing, breathing rocket. I guess they want to take part in the same risk as we three have to at that point in the launch sequence.

At this point, though, they've moved everybody away. The ride is incredible. I don't know how to describe it. There's a lot of rumbling noise, vibration. Very abrupt cutoffs as we go through staging, and then there's peace and quiet when you get to orbit. And all the hoopla you went through getting to the launch pad is kind of behind you.

You think about, if you have a reflective moment, your family back at the launch pad, thousands of miles away already.

If you were to watch the faces of launch teams at Cape Canaveral, and the managers, you would find expressions of concern and nervousness and prayer and hope written all over their faces. At this moment, if people have forgotten the risk of the launch, then they remember it.

On board, it's more simple. Crew members have to only perform reliably and carefully. In my mind, once embarked on a risky phase, be it crossing a crevice field on a glacier or carrying out procedures using dynamic operations in a space vehicle, at that point, you have to stop worrying and move on to minimize the risk of your own failure. That's the risk-taker's point of view.

Of course, there's time to relax sometimes, such as New Year. A long-duration mission is very much an act of endurance and perseverance. The risk I take most seriously is being part of a crew that cannot shift out of relaxation from routine to operational readiness for dynamic operations. An example of that would be shifting to operational readiness for reentry in a space vehicle after you've spent 194 days in space.

This transition for our crew, including a long-time unseen flight engineer, was probably the greatest risk we were exposed to during this otherwise pretty nominal expedition. The ride is incredible. From four hours ago, [when] we were enjoying chocolate and drinks, and then, after a deorbit burn, pyro belts firing, tumbling, the shock of parachute opening, rapid depressurization of the spacecraft, and then the smell of cordite coming in through the vents of the spacecraft into the cabin; finally, you touch down onto the Kazakhstan plain.

[video] He's saying, "I congratulate you." This is the hole made by a thruster made on the Soyuz spacecraft as it did the braking burn.

After the risk is past, crew members, family, space managers, all of us are relieved, and we celebrate how we have cheated death once more. It shows in our faces that the risk of spaceflight and space exploration is always present, and we must always be honest about it, explain it, and do our utmost to reduce it, without hiding it. That way, when we risk-takers are back with our families and we talk about committing to new space exploration—she says, "Don't you dare fly again!" [joking] No, you talk about it. Nobody should ever, ever be shocked if, in taking those steps, we should falter and not return home.

Exploration today carries risk just as dangerous as it did in history. I believe we must honestly explain that risk, just as we move forward to carry out the

President's space exploration vision. Americans can suffer discomfort, hardship, and overcome the greatest difficulties when the goals and risks are laid out plainly side by side. We must take on these most challenging adventures, while looking into the face of risk. In that way, we will achieve some incredible things in space.

You've listened this evening to me and the excellent discussion today.

Please continue to let me and us know what you, the public, and our Congress, think about risk-taking in space exploration. Thank you for being here this evening.

About the Editors

Steven J. Dick is the chief historian for NASA. He has a B.S. in astrophysics (1971), and an M.A. and Ph.D. (1977) in history and philosophy of science from Indiana University. Dick worked as an astronomer and historian of science at the U. S. Naval Observatory in Washington, D.C. for 24 years before coming to NASA Headquarters in 2003. Among his books are *Plurality of Worlds: The Origins of the Extraterrestrial Life Debate from Democritus to Kant* (1982), *The Biological Universe: The Twentieth Century Extraterrestrial Life Debate and the Limits of Science* (1996), and *Life on Other Worlds* (1998), the latter translated into Chinese, Italian, Czech, and Polish. His most recent books are *The Living Universe: NASA and the Development of Astrobiology* (2004) and a comprehensive history of the U. S. Naval Observatory, *Sky and Ocean Joined: The U. S. Naval Observatory, 1830–2000* (2003). The latter received the Pendleton Prize of the Society for History in the Federal Government. He is also editor of *Many Worlds: The New Universe, Extraterrestrial Life and the Theological Implications* (2000). Dick is the recipient of the Navy Meritorious Civilian Service Medal and the NASA Group Achievement Award for his role in NASA's multidisciplinary program in astrobiology. He has served as chairman of the Historical Astronomy Division of the American Astronomical Society, as president of the History of Astronomy Commission of the International Astronomical Union, and is the immediate past president of the Philosophical Society of Washington.

Keith L. Cowing is the editor of the online publications *NASA Watch* and *SpaceRef.com*. He received both a B.A. (1984) and M.A. (1987) in biology from Central Connecticut State University. Cowing is the coauthor (with Frank Sietzen, Jr.) of *New Moon Rising* (2004) which examines the events leading up to the formulation of the Vision for Space Exploration as put forth by President Bush. Cowing has served as guest editor of *Ad Astra Magazine*, written for United Press International, and *Air&Space Magazine* (among others), and appeared on all major U.S. television and radio networks as a commentator on space policy and technology. Prior to being an online journalist, Cowing was a NASA civil servant and served as manager of Pressurized Payload Accommodations at the NASA Space Station Freedom Program Office. Prior to, and immediately after, working for NASA, Cowing was employed by the American Institute of Biological Sciences where he managed a series of large biomedical peer review activities for NASA and the U.S. Army. Cowing has been a participant in the NASA Haughton Mars Project located on Devon Island, Nunavut, Canada. Together with his business partner Marc Boucher, SpaceRef Interactive donated and constructed the Arthur Clarke Mars Greenhouse, now in operating on Devon Island. Cowing was also part of the early organizational meetings that led to the establishment of NASA's astrobiology program and the NASA Astrobiology Institute. Cowing helped to organize the symposium and donated his services in the preparation of these proceedings.

Index

A

M

S

The NASA History Series

REFERENCE WORKS, NASA SP-4000:

Grimwood, James M. *Project Mercury: A Chronology.* NASA SP-4001, 1963.

Grimwood, James M., and C. Barton Hacker, with Peter J. Vorzimmer. *Project Gemini Technology and Operations: A Chronology.* NASA SP-4002, 1969.

Link, Mae Mills. *Space Medicine in Project Mercury.* NASA SP-4003, 1965.

Astronautics and Aeronautics, 1963: Chronology of Science, Technology, and Policy. NASA SP-4004, 1964.

Astronautics and Aeronautics, 1964: Chronology of Science, Technology, and Policy. NASA SP-4005, 1965.

Astronautics and Aeronautics, 1965: Chronology of Science, Technology, and Policy. NASA SP-4006, 1966.

Astronautics and Aeronautics, 1966: Chronology of Science, Technology, and Policy. NASA SP-4007, 1967.

Astronautics and Aeronautics, 1967: Chronology of Science, Technology, and Policy. NASA SP-4008, 1968.

Ertel, Ivan D., and Mary Louise Morse. *The Apollo Spacecraft: A Chronology, Volume I, Through November 7, 1962.* NASA SP-4009, 1969.

Morse, Mary Louise, and Jean Kernahan Bays. *The Apollo Spacecraft: A Chronology, Volume II, November 8, 1962–September 30, 1964.* NASA SP-4009, 1973.

Brooks, Courtney G., and Ivan D. Ertel. *The Apollo Spacecraft: A Chronology, Volume III, October 1, 1964–January 20, 1966.* NASA SP-4009, 1973.

Ertel, Ivan D., and Roland W. Newkirk, with Courtney G. Brooks. *The Apollo Spacecraft: A Chronology, Volume IV, January 21, 1966–July 13, 1974.* NASA SP-4009, 1978.

Astronautics and Aeronautics, 1968: Chronology of Science, Technology, and Policy. NASA SP-4010, 1969.

Newkirk, Roland W., and Ivan D. Ertel, with Courtney G. Brooks. *Skylab: A Chronology.* NASA SP-4011, 1977.

Van Nimmen, Jane, and Leonard C. Bruno, with Robert L. Rosholt. *NASA Historical Data Book, Volume I: NASA Resources, 1958–1968.* NASA SP-4012, 1976, rep. ed. 1988.

Ezell, Linda Neuman. *NASA Historical Data Book, Volume II: Programs and Projects, 1958–1968.* NASA SP-4012, 1988.

Ezell, Linda Neuman. *NASA Historical Data Book, Volume III: Programs and Projects, 1969–1978*. NASA SP-4012, 1988.

Gawdiak, Ihor Y., with Helen Fedor, compilers. *NASA Historical Data Book, Volume IV: NASA Resources, 1969–1978*. NASA SP-4012, 1994.

Rumerman, Judy A., compiler. *NASA Historical Data Book, 1979–1988: Volume V, NASA Launch Systems, Space Transportation, Human Spaceflight, and Space Science*. NASA SP-4012, 1999.

Rumerman, Judy A., compiler. *NASA Historical Data Book, Volume VI: NASA Space Applications, Aeronautics and Space Research and Technology, Tracking and Data Acquisition/Space Operations, Commercial Programs, and Resources, 1979–1988*. NASA SP-2000-4012, 2000.

Astronautics and Aeronautics, 1969: Chronology of Science, Technology, and Policy. NASA SP-4014, 1970.

Astronautics and Aeronautics, 1970: Chronology of Science, Technology, and Policy. NASA SP-4015, 1972.

Astronautics and Aeronautics, 1971: Chronology of Science, Technology, and Policy. NASA SP-4016, 1972.

Astronautics and Aeronautics, 1972: Chronology of Science, Technology, and Policy. NASA SP-4017, 1974.

Astronautics and Aeronautics, 1973: Chronology of Science, Technology, and Policy. NASA SP-4018, 1975.

Astronautics and Aeronautics, 1974: Chronology of Science, Technology, and Policy. NASA SP-4019, 1977.

Astronautics and Aeronautics, 1975: Chronology of Science, Technology, and Policy. NASA SP-4020, 1979.

Astronautics and Aeronautics, 1976: Chronology of Science, Technology, and Policy. NASA SP-4021, 1984.

Astronautics and Aeronautics, 1977: Chronology of Science, Technology, and Policy. NASA SP-4022, 1986.

Astronautics and Aeronautics, 1978: Chronology of Science, Technology, and Policy. NASA SP-4023, 1986.

Astronautics and Aeronautics, 1979–1984: Chronology of Science, Technology, and Policy. NASA SP-4024, 1988.

Astronautics and Aeronautics, 1985: Chronology of Science, Technology, and Policy. NASA SP-4025, 1990.

Noordung, Hermann. *The Problem of Space Travel: The Rocket Motor*. Edited by Ernst Stuhlinger and J. D. Hunley, with Jennifer Garland. NASA SP-4026, 1995.

Astronautics and Aeronautics, 1986–1990: A Chronology. NASA SP-4027, 1997.

Astronautics and Aeronautics, 1990–1995: A Chronology. NASA SP-2000-4028, 2000.

MANAGEMENT HISTORIES, NASA SP-4100:

Rosholt, Robert L. *An Administrative History of NASA, 1958–1963*. NASA SP-4101, 1966.

Levine, Arnold S. *Managing NASA in the Apollo Era*. NASA SP-4102, 1982.

Roland, Alex. *Model Research: The National Advisory Committee for Aeronautics, 1915–1958*. NASA SP-4103, 1985.

Fries, Sylvia D. *NASA Engineers and the Age of Apollo*. NASA SP-4104, 1992.

Glennan, T. Keith. *The Birth of NASA: The Diary of T. Keith Glennan*. J. D. Hunley, editor. NASA SP-4105, 1993.

Seamans, Robert C., Jr. *Aiming at Targets: The Autobiography of Robert C. Seamans, Jr.* NASA SP-4106, 1996.

Garber, Stephen J., editor. *Looking Backward, Looking Forward: Forty Years of U.S. Human Spaceflight Symposium*. NASA SP-2002-4107, 2002.

Chertok, Boris. *Rockets and People, Volume I*. NASA-SP-2005-4110, 2005.

Laufer, Dr. Alexander, Todd Post, and Dr. Edward J. Hoffman. *Shared Voyage: Learning and Unlearning from Remarkable Projects*. NASA-SP-2005-4111, 2005.

PROJECT HISTORIES, NASA SP-4200:

Swenson, Loyd S., Jr., James M. Grimwood, and Charles C. Alexander. *This New Ocean: A History of Project Mercury*. NASA SP-4201, 1966; rep. ed. 1998.

Green, Constance McLaughlin, and Milton Lomask. *Vanguard: A History*. NASA SP-4202, 1970; rep. ed. Smithsonian Institution Press, 1971.

Hacker, Barton C., and James M. Grimwood. *On the Shoulders of Titans: A History of Project Gemini*. NASA SP-4203, 1977.

Benson, Charles D., and William Barnaby Faherty. *Moonport: A History of Apollo Launch Facilities and Operations*. NASA SP-4204, 1978.

Brooks, Courtney G., James M. Grimwood, and Loyd S. Swenson, Jr. *Chariots for Apollo: A History of Manned Lunar Spacecraft*. NASA SP-4205, 1979.

Bilstein, Roger E. *Stages to Saturn: A Technological History of the Apollo/Saturn Launch Vehicles*. NASA SP-4206, 1980, rep. ed. 1997.

SP-4207 not published.

Compton, W. David, and Charles D. Benson. *Living and Working in Space: A History of Skylab*. NASA SP-4208, 1983.

Ezell, Edward Clinton, and Linda Neuman Ezell. *The Partnership: A History of the Apollo-Soyuz Test Project.* NASA SP-4209, 1978.

Hall, R. Cargill. *Lunar Impact: A History of Project Ranger.* NASA SP-4210, 1977.

Newell, Homer E. *Beyond the Atmosphere: Early Years of Space Science.* NASA SP-4211, 1980.

Ezell, Edward Clinton, and Linda Neuman Ezell. *On Mars: Exploration of the Red Planet, 1958–1978.* NASA SP-4212, 1984.

Pitts, John A. *The Human Factor: Biomedicine in the Manned Space Program to 1980.* NASA SP-4213, 1985.

Compton, W. David. *Where No Man Has Gone Before: A History of Apollo Lunar Exploration Missions.* NASA SP-4214, 1989.

Naugle, John E. *First Among Equals: The Selection of NASA Space Science Experiments.* NASA SP-4215, 1991.

Wallace, Lane E. *Airborne Trailblazer: Two Decades with NASA Langley's Boeing 737 Flying Laboratory.* NASA SP-4216, 1994.

Butrica, Andrew J., editor. *Beyond the Ionosphere: Fifty Years of Satellite Communication.* NASA SP-4217, 1997.

Butrica, Andrew J. *To See the Unseen: A History of Planetary Radar Astronomy.* NASA SP-4218, 1996.

Mack, Pamela E., editor. *From Engineering Science to Big Science: The NACA and NASA Collier Trophy Research Project Winners.* NASA SP-4219, 1998.

Reed, R. Dale, with Darlene Lister. *Wingless Flight: The Lifting Body Story.* NASA SP-4220, 1997.

Heppenheimer, T. A. *The Space Shuttle Decision: NASA's Search for a Reusable Space Vehicle.* NASA SP-4221, 1999.

Hunley, J. D., editor. *Toward Mach 2: The Douglas D-558 Program.* NASA SP-4222, 1999.

Swanson, Glen E., editor. *"Before this Decade Is Out . . .": Personal Reflections on the Apollo Program.* NASA SP-4223, 1999.

Tomayko, James E. *Computers Take Flight: A History of NASA's Pioneering Digital Fly-by-Wire Project.* NASA SP-2000-4224, 2000.

Morgan, Clay. *Shuttle-Mir: The U.S. and Russia Share History's Highest Stage.* NASA SP-2001-4225, 2001.

Leary, William M. *"We Freeze to Please": A History of NASA's Icing Research Tunnel and the Quest for Flight Safety.* NASA SP-2002-4226, 2002.

Mudgway, Douglas J. *Uplink-Downlink: A History of the Deep Space Network 1957–1997.* NASA SP-2001-4227, 2001.

CENTER HISTORIES, NASA SP-4300:

Rosenthal, Alfred. *Venture into Space: Early Years of Goddard Space Flight Center.* NASA SP-4301, 1985.

Hartman, Edwin P. *Adventures in Research: A History of Ames Research Center, 1940–1965.* NASA SP-4302, 1970.

Hallion, Richard P. *On the Frontier: Flight Research at Dryden, 1946–1981.* NASA SP-4303, 1984.

Muenger, Elizabeth A. *Searching the Horizon: A History of Ames Research Center, 1940–1976.* NASA SP-4304, 1985.

Hansen, James R. *Engineer in Charge: A History of the Langley Aeronautical Laboratory, 1917–1958.* NASA SP-4305, 1987.

Dawson, Virginia P. *Engines and Innovation: Lewis Laboratory and American Propulsion Technology.* NASA SP-4306, 1991.

Dethloff, Henry C. *"Suddenly Tomorrow Came . . .": A History of the Johnson Space Center.* NASA SP-4307, 1993.

Hansen, James R. *Spaceflight Revolution: NASA Langley Research Center from Sputnik to Apollo.* NASA SP-4308, 1995.

Wallace, Lane E. *Flights of Discovery: 50 Years at the NASA Dryden Flight Research Center.* NASA SP-4309, 1996.

Herring, Mack R. *Way Station to Space: A History of the John C. Stennis Space Center.* NASA SP-4310, 1997.

Wallace, Harold D., Jr. *Wallops Station and the Creation of the American Space Program.* NASA SP-4311, 1997.

Wallace, Lane E. *Dreams, Hopes, Realities: NASA's Goddard Space Flight Center, The First Forty Years.* NASA SP-4312, 1999.

Dunar, Andrew J., and Stephen P. Waring. *Power to Explore: A History of the Marshall Space Flight Center.* NASA SP-4313, 1999.

Bugos, Glenn E. *Atmosphere of Freedom: Sixty Years at the NASA Ames Research Center.* NASA SP-2000-4314, 2000.

GENERAL HISTORIES, NASA SP-4400:

Corliss, William R. *NASA Sounding Rockets, 1958–1968: A Historical Summary.* NASA SP-4401, 1971.

Wells, Helen T., Susan H. Whiteley, and Carrie Karegeannes. *Origins of NASA Names.* NASA SP-4402, 1976.

Anderson, Frank W., Jr. *Orders of Magnitude: A History of NACA and NASA, 1915–1980.* NASA SP-4403, 1981.

Sloop, John L. *Liquid Hydrogen as a Propulsion Fuel, 1945–1959.* NASA SP-4404, 1978.

Roland, Alex. *A Spacefaring People: Perspectives on Early Spaceflight.* NASA SP-4405, 1985.

Bilstein, Roger E. *Orders of Magnitude: A History of the NACA and NASA, 1915–1990.* NASA SP-4406, 1989.

Logsdon, John M., editor, with Linda J. Lear, Jannelle Warren-Findley, Ray A. Williamson, and Dwayne A. Day. *Exploring the Unknown: Selected Documents in the History of the U.S. Civil Space Program, Volume I, Organizing for Exploration.* NASA SP-4407, 1995.

Logsdon, John M., editor, with Dwayne A. Day and Roger D. Launius. *Exploring the Unknown: Selected Documents in the History of the U.S. Civil Space Program, Volume II, Relations with Other Organizations.* NASA SP-4407, 1996.

Logsdon, John M., editor, with Roger D. Launius, David H. Onkst, and Stephen J. Garber. *Exploring the Unknown: Selected Documents in the History of the U.S. Civil Space Program, Volume III, Using Space.* NASA SP-4407, 1998.

Logsdon, John M., general editor, with Ray A. Williamson, Roger D. Launius, Russell J. Acker, Stephen J. Garber, and Jonathan L. Friedman. *Exploring the Unknown: Selected Documents in the History of the U.S. Civil Space Program, Volume IV, Accessing Space.* NASA SP-4407, 1999.

Logsdon, John M., general editor, with Amy Paige Snyder, Roger D. Launius, Stephen J. Garber, and Regan Anne Newport. *Exploring the Unknown: Selected Documents in the History of the U.S. Civil Space Program, Volume V, Exploring the Cosmos.* NASA SP-2001-4407, 2001.

Siddiqi, Asif A. *Challenge to Apollo: The Soviet Union and the Space Race, 1945–1974.* NASA SP-2000-4408, 2000.

MONOGRAPHS IN AEROSPACE HISTORY, NASA SP-4500:

Launius, Roger D. and Aaron K. Gillette, compilers, *Toward a History of the Space Shuttle: An Annotated Bibliography.* Monograph in Aerospace History, No. 1, 1992.

Launius, Roger D., and J. D. Hunley, compilers, *An Annotated Bibliography of the Apollo Program.* Monograph in Aerospace History, No. 2, 1994.

Launius, Roger D. *Apollo: A Retrospective Analysis.* Monograph in Aerospace History, No. 3, 1994.

Hansen, James R. *Enchanted Rendezvous: John C. Houbolt and the Genesis of the Lunar-Orbit Rendezvous Concept.* Monograph in Aerospace History, No. 4, 1995.

Gorn, Michael H. *Hugh L. Dryden's Career in Aviation and Space.* Monograph in Aerospace History, No. 5, 1996.

Powers, Sheryll Goecke. *Women in Flight Research at NASA Dryden Flight Research Center, from 1946 to 1995.* Monograph in Aerospace History, No. 6, 1997.

Portree, David S. F. and Robert C. Trevino. *Walking to Olympus: An EVA Chronology.* Monograph in Aerospace History, No. 7, 1997.

Logsdon, John M., moderator. *Legislative Origins of the National Aeronautics and Space Act of 1958: Proceedings of an Oral History Workshop.* Monograph in Aerospace History, No. 8, 1998.

Rumerman, Judy A., compiler, *U.S. Human Spaceflight, A Record of Achievement 1961–1998.* Monograph in Aerospace History, No. 9, 1998.

Portree, David S. F. *NASA's Origins and the Dawn of the Space Age.* Monograph in Aerospace History, No. 10, 1998.

Logsdon, John M. *Together in Orbit: The Origins of International Cooperation in the Space Station.* Monograph in Aerospace History, No. 11, 1998.

Phillips, W. Hewitt. *Journey in Aeronautical Research: A Career at NASA Langley Research Center.* Monograph in Aerospace History, No. 12, 1998.

Braslow, Albert L. *A History of Suction-Type Laminar-Flow Control with Emphasis on Flight Research.* Monograph in Aerospace History, No. 13, 1999.

Logsdon, John M., moderator. *Managing the Moon Program: Lessons Learned From Apollo.* Monograph in Aerospace History, No. 14, 1999.

Perminov, V. G. *The Difficult Road to Mars: A Brief History of Mars Exploration in the Soviet Union.* Monograph in Aerospace History, No. 15, 1999.

Tucker, Tom. *Touchdown: The Development of Propulsion Controlled Aircraft at NASA Dryden.* Monograph in Aerospace History, No. 16, 1999.

Maisel, Martin D., Demo J. Giulianetti, and Daniel C. Dugan. *The History of the XV-15 Tilt Rotor Research Aircraft: From Concept to Flight.* NASA SP-2000-4517, 2000.

Jenkins, Dennis R. *Hypersonics Before the Shuttle: A Concise History of the X-15 Research Airplane.* NASA SP-2000-4518, 2000.

Chambers, Joseph R. *Partners in Freedom: Contributions of the Langley Research Center to U.S. Military Aircraft in the 1990s.* NASA SP-2000-4519, 2000.

Waltman, Gene L. *Black Magic and Gremlins: Analog Flight Simulations at NASA's Flight Research Center.* NASA SP-2000-4520, 2000.

Portree, David S. F. *Humans to Mars: Fifty Years of Mission Planning, 1950–2000.* NASA SP-2001-4521, 2001.

Thompson, Milton O., with J. D. Hunley. *Flight Research: Problems Encountered and What They Should Teach Us.* NASA SP-2000-4522, 2000.

Tucker, Tom. *The Eclipse Project.* NASA SP-2000-4523, 2000.

Siddiqi, Asif A. *Deep Space Chronicle: A Chronology of Deep Space and Planetary Probes,* 1958–2000. NASA SP-2002-4524, 2002.

Merlin, Peter W. *Mach 3+: NASA/USAF YF-12 Flight Research, 1969–1979.* NASA SP-2001-4525, 2001.

Anderson, Seth B. *Memoirs of an Aeronautical Engineer—Flight Tests at Ames Research Center: 1940–1970.* NASA SP-2002-4526, 2002.

Renstrom, Arthur G. *Wilbur and Orville Wright: A Bibliography Commemorating the One-Hundredth Anniversary of the First Powered Flight on December 17, 1903.* NASA SP-2002-4527, 2002.

No monograph 28.

Chambers, Joseph R. *Concept to Reality: Contributions of the NASA Langley Research Center to U.S. Civil Aircraft of the 1990s.* SP-2003-4529, 2003.

Peebles, Curtis, editor. *The Spoken Word: Recollections of Dryden History, The Early Years.* SP-2003-4530, 2003.

Jenkins, Dennis R., Tony Landis, and Jay Miller. *American X-Vehicles: An Inventory—X-1 to X-50.* SP-2003-4531, 2003.

Renstrom, Arthur G. *Wilbur and Orville Wright Chronology.* Monographs in Aerospace History, No. 32, 2003. NASA SP-2003-4532.

Bowles, Mark D., and Robert S. Arrighi. NASA's *Nuclear Frontier: The Plum Brook Reactor Facility, 1941–2002.* Monographs in Aerospace History, No. 33, 2004. NASA SP-2004-4533.

McCurdy, Howard E. *Low-Cost Innovation in Spaceflight.* Monographs in Aerospace History, No. 36, 2005. NASA SP-2005-4536.

Seamans, Robert C. *Project Apollo: The Tough Decisions.* Monographs in Aerospace History, No. 37, 2005. NASA SP-2005-4537.

Lambright, Henry W. *NASA and the Environment. The Case of Ozone Depletion.* Monographs in Aerospace History, No. 38, 2005. NASA SP-2005-4538.